Microregional Fragmentation

Contributions to Economics

Albrecht Ritschl
Prices and Production
– Elements of a System – Theoretic Perspective –
1989. 159 pp. Softcover DM 59,–
ISBN 3-7908-0429-0

Arnulf Grübler
The Rise and Fall of Infrastructures
– Dynamics of Evolution and Technological Change in Transport –
1990. 305 pp. Softcover DM 85,–
ISBN 3-7908-0479-7

Peter R. Haiss
Cultural Influences on Strategic Planning
1990. 188 pp. Softcover DM 65,–
ISBN 3-7908-0481-9

Manfred Kremer/Marion Weber (Eds.)
Transforming Economic Systems: The Case of Poland
1992. 179 pp. Softcover DM 69,–
ISBN 3-7908-91415-3

Marcel F. van Marion
Liberal Trade and Japan
1993. 298 pp. Softcover DM 90,–
ISBN 3-7908-0699-4

Michael Carlberg
Open Economy Dynamics
1993. 203 pp. Softcover DM 75,–
ISBN 3-7908-0708-7

Hans Schneeweiß/
Klaus F. Zimmermann (Eds.)
Studies in Applied Econometrics
1993. 238 pp. Softcover DM 85,–
ISBN 3-7908-0716-8

Gerhard Gehrig/Wladyslaw Welfe (Eds.)
Economies in Transition
1993. 292 pp. Softcover DM 90,–
ISBN 3-7908-0721-4

Alfred Franz/Carsten Stahmer (Eds)
Approaches to Environmental Accounting
1993. 542 pp. Softcover DM 178,–
ISBN 3-7908-0719-2

János Gács / Georg Winckler (Eds.)
International Trade and Restructuring in Eastern Europe
1994. 343 pp. Softcover DM 98,–
ISBN 3-7908-0759-1

Christoph M. Schneider
Research and Development Management: From the Soviet Union to Russia
1994. 253 pp. Softcover DM 85,–
ISBN 3-7908-0757-5

Valeria De Bonis
Stabilization Policy in an Exchange Rate Union
1994. 172 pp. Softcover DM 75,–
ISBN 3-7908-0789-3

Bernhard Böhm / Lionello F. Punzo (Eds.)
Economic Performance
1994. 323 pp. Softcover DM 98,–
ISBN 3-7908-0811-3

Karl Steininger
Trade and Environment
1995. 219 pp. Softcover DM 75,–
ISBN 3-7908-0814-8

Michael Reiter
The Dynamics of Business Cycles
1995. 215 pp. Softcover DM 75,–
ISBN 3-7908-0823-7

Michael Carlberg
Sustainability and Optimality of Public Debt
1995. 217 pp. Softcover DM 85,–
ISBN 3-7908-0834-2

Lars Olof Persson · Ulf Wiberg

Microregional Fragmentation

Contrasts Between a Welfare State
and a Market Economy

With 47 Figures

Physica-Verlag
A Springer-Verlag Company

Series Editors
Werner A. Müller
Peter Schuster

Authors
Lars Olof Persson
Senior Researcher
Research Group on Regional Analysis (FORA)
The Royal Institute of Technology
S-100 44 Stockholm, Sweden

Ulf Wiberg
Associate Professor
Centre for Regional Science (CERUM)
Umeå University
S-901 87 Umeå, Sweden

ISBN 3-7908-0855-5 Physica-Verlag Heidelberg

Die Deutsche Bibliothek - CIP-Einheitsaufnahme
Persson, Lars Olof:
Microregional fragmentation: contrasts between a welfare state
and a market economy / Lars Olof Persson; Ulf Wiberg. –
Heidelberg: Physica-Verl., 1995
(Contributions to economics)
ISBN 3-7908-0855-5
NE: Wiberg, Ulf:

This work is subject to copyright. All rights are reserved, whether the whole or part of the material is concerned, specifically the rights of translation, reprinting, reuse of illustration, recitation, broadcasting, reproduction on microfilms or in other ways, and storage in data banks. Duplication of this publication or parts thereof is only permitted under the provisions of the German Copyright Law of September 9, 1965, in its version of June 24, 1985, and a copyright fee must always be paid. Violations fall under the prosecution act of the German Copyright Law.

©Physica-Verlag Heidelberg 1995
Printed in Germany

The use of registered names, trademarks, etc. in this publication does not imply, even in the absence of a specific statement, that such names are exempt from the relevant protective laws and regulations and therefore free for general use.

88/2202-5 4 3 2 1 0 – Printed on acid-free paper

PREFACE

This book is written primarily for a Scandinavian and European audience interested in regional policy and planning. Attention is placed on the transformation process in the Swedish economy and its implications for regional balances of socio-economic conditions and changes in spatial structures. Conditions in the United States, especially North Carolina, are used as a reference.

The book is based on work originating within the framework of an international forum for exchange of ideas and co-operation between researchers, planners and practitioners, *The Consortium for the Study of Perceived Planning Issues in Marginal Areas - PIMA*. The group was established in 1989 and is interested in various aspects of marginal areas defined either in locational or developmental terms. Members of the core group represent universities in the United States, Sweden and Ireland.

During recent years a subgroup within PIMA has focused attention on studies of areas located between urban centres and rural peripheries. These areas have been labelled Intermediate Socio-economic Regions - ISER. Joint work between Sweden and North Carolina of a comparative nature has been conducted by the authors of this book and Professor Ole Gade and some of his students at Appalachian State University, North Carolina. This work has been published in proceedings from PIMA meetings (*Planning Issues in Marginal Areas*, Boone: Ole Gade, Vincent P. Miller Jr. and Lawrence M. Sommers, eds. 1991; *Planning and Development of Marginal Areas*, Galway: Mícheál O'Cinneide and Seamus Grimes, eds. 1992; *Spatial Dynamics of Highland and High Latitude Environments Symposium Proceedings*, Boone: Ole Gade, ed. 1992; and *Marginal Areas in Developed Countries*, Umeå: Ulf Wiberg, ed. 1993). We draw heavily from the contributions of Gade and Jones in these proceedings and other sources. Vital sections of chapter 4 are derived from Sten Axelsson, Svante Berglund and Lars Olof Persson (1994) with the kind permission of the co-authors.

We would like to give special thanks to Ole Gade for the field work in both Sweden and North Carolina, and for an interesting exchange and elaboration of ideas. Jeff A. Jones has also been of great help and stayed with us in Sweden for some months to prepare his contribution to our project. We also appreciate the work done by Erik Sondell and Kenneth Ennefors in collecting and handling Swedish empirical material. Seminars within PIMA have provided us with excellent opportunities to present and receive comments on our research. We have also from Dr. Stephen Fournier, Department of Infrastructure and Planning, The Royal Institute of Technology, Stockholm, received valuable comments and help with the editing of an earlier manuscript. Erik Sondell has made the final preparation of the manuscript for publication. In the final procedure has also Jennifer Wundersitz been helpful with English language editing and Marika Hedlund has made a major part of the graphics generation.

The project has been supported by the Swedish Council for Building Research, the Expert Group on Regional and Urban Studies (ERU) within the Ministry of Labour, Sweden and the Centre for Regional Science (CERUM), Umeå University.

Stockholm and Umeå, January 1995

Lars Olof Persson   Ulf Wiberg

CONTENTS

1 **Dissolution of the Socio-Economic Conformity** 1
   1.1 Introduction 1
   1.2 Objectives of the Study 2
   1.3 Spatial Differences in Socio-Economic Development 3
   1.4 The Quality of Life 6

2 **Processes Reshaping the Spatial Structure** 13
   2.1 Introduction 13
   2.2 From a Resource-Based to a Network-Based Economy 13
   2.3 Quality of Life Improvements 18
   2.4 Emerging Spatial Structures 22
   2.5 Function and Quality of the ISER in Sweden 24

3 **Patterns of Sectoral and Spatial Change** 25
   3.1 Internationalisation and Locality 25
   3.2 Social Change and Preferences 33
   3.3 Distribution of Income and Services 39
   3.4 Focus on Intermediate Regions in the Swedish Geography 45

4 **Spatial Dimensions of the Emerging Knowledge Society in Sweden** 51
   4.1 The Race for Increased Productivity 51
   4.2 Characteristics of the Knowledge Society 58
   4.3 Fragmentation Tendencies Among Swedish
       Local Labour Markets 62
   4.4 Patterns of Restructuring Labour Market Areas 67

	4.5 Migration Patterns	72
	4.6 Conclusions on Mobility in Sweden	85
5	**Regions and Contexts**	87
	5.1 Comparing Two Contexts	87
	5.2 Socio-Economic Regions in Two Countries	88
	5.3 Sweden in the European Union	96
6	**Policy and Planning Perspectives**	99
	6.1 Introduction	99
	6.2 Sweden 1930 - 1990: The Central Role of the Welfare State	99
	6.3 The Intermediate Role of Municipal Planning	102
	6.4 Approaching a Bottom-Up Planning Model	103
	6.5 Alternative Planning Modes	104
	6.6 Policy and Planning in the United States	106
	6.7 Sweden in the First Half of the 1990s	109
7	**Sweden Facing a New Micro- and Macroregional Fragmentation**	115
	7.1 Uniformity and Fragmentation	115
	7.2 Four Carriers of Change	117
	7.3 Strategic Issues	120
	References	123
	Author Index	129
	Subject Index	131

1 DISSOLUTION OF THE SOCIO-ECONOMIC CONFORMITY

1.1 Introduction

In Scandinavian countries, processes of economic integration, political supernationalism and the globalization of investment, are shaking the traditional foundations of strongly centralised national governments and guided regional and urban-rural development. Scandinavians are finding that the increasing costs of maintaining regional equities in quality of life conditions are threatening their ability to compete effectively in unrestricted world markets. Their response is an increasing tendency towards accepting the standards of private enterprise, free market competition and relatively unfettered capitalistic development. This, they realise, must come in part through a reduction in welfare and regional transfer payments, and with a concomitant decentralisation of governmental authority. What this brings with it is an increasing privatisation of their economies accompanied by an emergence of individual and corporate dominance in the localisation of economic activity. The result is a potential for a new economic structure that, unencumbered by regional development directives, may have unpredictable land use implications.

Will these changes in Scandinavian life lead to the evolution of new settlement patterns? What will be the implications for traditional community and regional development? Will the traditional egalitarian philosophy that influenced the creation of regions and communities with equal opportunities and amenities gradually give away to laissez-faire and individualism? Will we see an "Americanisation" of the Scandinavian landscape with a move toward increasingly dichotomised development?[1]

---

[1] The further discussion in this chapter is largely abstracted from Gade, Persson and Wiberg (1992).

1.2 Objectives of the Study

We will focus on the dynamics of what we call "The Intermediate Socio-Economic Region" (ISER), which means areas located between cities and rural peripheries, where we find a variety of mixed urban and rural elements. The first rationale for focusing on these regions is that we expect that the urban-rural mix has an increasing power of attraction. The second rationale is that we expect ISERs to remain important links in the settlement structure, especially with respect to their support for remote areas. We will discuss how this region of urbanizing rural space comes into being, and what its essential conditions of life and livelihood are. We also will demonstrate the implications of this evolution on existing patterns of land use, labour mobility, and population settlement. We further believe that the changing American processes of contemporary land use have extraordinary implications for a European public increasingly persuaded that the American approach to governance and economic life is the key to improved competitiveness and lasting socio-economic health. So we seek to discover the degree to which American trends are moving toward being replicated in Sweden. We believe that an Americanisation of Swedish economic structures and activities will lead to an undesirable degree of local and regional dichotomisation of both opportunity and wealth.

In Sweden, where compensatory elements are embedded in almost all sectors of the welfare system, regional impacts are substantial and deep-rooted. We suggest, therefore, that an analysis of a number of different aspects of the welfare system is crucial to understanding recent and potential regional development. Towards the end of this century, we anticipate several changes in the Swedish welfare system, including an increasing influence of market forces in service production, and additional elements of decentralised problem solving and planning. These changes flow from shifts in international competition, budgetary problems at the national level, and shifting attitudes and perceptions of needs.

Our study has an explorative character. Several cluster analyses, taking into consideration various types of indicators, will be used in order to demonstrate the spatial dimension of economic and social conditions and future opportunities.

1.3 Spatial Differences in Socio-Economic Development

Practically all nations and regions - whether developed or not, industrialized or not, under a free market or a more or less planned regime - have settlement patterns that can be characterized by urban-rural continua, representing geographic differences in socio-economic development. It is quite likely that the slope of the urban-rural continuum will vary by stage of development in terms of materialistic quality of life, from a high in urban centres to a low in peripheral rural regions. We expect that the steeper the slope, the less developed will be national conditions, with the least slope likely to be found in advanced social welfare states, like Sweden. However, it is also true that in most cases of regional policy and planning, this continuity in socio-economic development is not well recognised. "Urban" and "rural" are often treated as two quite opposite concepts. In most contexts when we think 'urban' we visualise built-up areas marked by intensive economic activity, high population density, good accessibility but also congestion problems, and a relatively high but unevenly distributed quality of life with respect to material resources. Therefore, these areas have problems that are thought to be unique, and planning has proceeded accordingly.

When we think 'rural' we see a much more dispersed population with economic activities tied largely to the supply and quality of natural resources, and a comparatively low quality of life in the sense previously mentioned. On average, the economic conditions as well as the social dynamics of the more urban region exceeds that of the more rural region. Consequently, in highly urbanized nations and regions, changes and developments affecting metropolitan or city regions often gain more political attention than those occurring in other regions. In general, it is likely that public policies directed toward urban regions have a more proactive character, while policies toward rural areas are predominantly reactive, in essence, they are designed to compensate for perceived low quality of life in a wide sense. However, we wish to move the centre of the urban-rural argument from its characteristic dichotomy, with urban and rural places placed as antagonists, to a view where people, socio-economic activities, and institutions, as well as time and space are regarded as dependent on each other.

As a point of departure for engaging in this task it is relevant to consider briefly the theoretical underpinnings of the study. Probably the earliest theoretical framework for understanding geographic continuities in land use was developed by von Thünen (Grotewold, 1959). Von Thünen effectively linked agricultural production type and intensity to distance from urban centres (markets), and thereby evolved the first tangible urban-rural continuum. In this instance we see urban and rural space linked together by increasing land values as urban areas are approached. If we move beyond consideration of land use to characteristics of population settlement, then we must look at central place theory. Its initial formulation was provided by Christaller and was based on his research on the distribution of urban places in southern Germany (Berry and Parr, 1988). However, with our particular focus on the dynamics of contemporary changes in the landscape, considerations of regularities in the sizing and spacing of central places, this approach falls somewhat short. In part the reason for this is the prior assumption of uniformity in land use, in the distribution of farmsteads, and in the persistence of an urban hierarchical arrangement. It needs to be noted though that several concepts deriving from central place theory are quite valuable in understanding and in dealing with contemporary rural issues. These include the notion of population threshold which defines the lower market limit for the successful operation of a particular function; the notion of the range of a good, defining the normal distance that a prospective purchaser is willing to travel to acquire a particular good or service; and the idea of nested trade areas, that on the basis of hierarchically structured amenity needs, provide for the existence of spatially interlinked demand and supply.

We also point to two further shortcomings of central place theory. It has a fixation on the static conditions of form, as opposed to process and change over time, and there is no attention to the emerging roles of functions and interactions across complex networks. New organisational forms and technologies within manufacturing, transmitting and processing of information, business services and administration, as well as faster modes of transportation of people are creating a wide variety of linkage patterns, often long distance in nature. Traditional hierarchically determined contacts are replaced (or new ones developed) with more open and direct networking with a horizontal character. Nodes which act merely as relays are outdated. A consequence of this new pattern of interaction and exchange is that not only nodal but also linkage qualities to places other than the nearest higher level centre become important. Parr also has noted that central place theory can never be regarded as a general model of an urban system, though "it is extremely useful for understanding the increasingly market oriented nature of economic activity within modern economies" (Parr, 1987, p. 236).

The quality of life concept encompasses many dimensions including both materialistic and non-materialistic concepts. An often used (but rather simple) measure is income which, in principal, only illustrates potential for materialistic quality of life. Figure 1.1 demonstrates that, on average, the quality of life gradient declines gradually as distance increases from the metropolitan centre to the rural periphery. Again, we postulate the degree of slope to be less for Sweden than is likely for United States and countries in Central Europe. We hold that there are likely to be considerable ups and downs in the quality of life gradient, and that

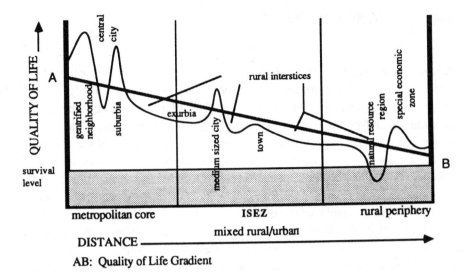

Fig. 1.1. A generalised view of the urban-rural quality of life gradient in the spatial context (after Chang, Gade and Jones, 1991).

these variations exist within any significant geographic portion of the socio-economic continuum represented here. Figure 1.1, for example, suggests that the relatively greatest gyrations occur where the quality of life condition is supposed to be highest to begin with, that is, within the metropolitan region. This is certainly the case in the United States, where we see clear evidence of the urban "hollowing out" process (Jackson, Hewings and Sonis, 1989), and is likely to be increasingly the case in Sweden. Recent decades of foreign immigration has resulted in ethnic segregation in the suburbs of metropolitan regions, with some housing areas showing poor socioeconomic characteristics. On the other hand, we also postulate that swings in the quality of life may occur within the peripheral rural regions. Here these undulations are largely related to the existence of specialised economic growth nodes, as, for example, tourism destination regions, centres of higher education, and areas of natural resource extraction.

1.4 The Quality of Life

A major objective in our study is to understand how international, national and regional transformation processes affect the quality of life in the ISERs. During the 1990s, the Swedish ISERs, traditionally influenced strongly by public decisions, will face more open market effects due both to a weakened public sector and to European integration and adjustment processes. In the US these areas traditionally have been exposed to immediate consequences of market changes. Conditions in the US therefore seem to provide meaningful references for probable future change in the Swedish ISERs. For example, we note the recent US manufacturing employer response to economic restructuring since the 1960s. Angel and Mitchell (1991) have documented that employers in a variety of industries have responded to increasing international competition and falling profits by reducing their direct labour costs. This has involved relocating the place of production from high to low-wage labour markets (frequently out of the country); subcontracting tasks to low wage employers; negotiating two tier employment contracts; replacing union labour with low wage non-union workers; and granting minimal wage increases, if not actual decreases. While the resulting regional disparities are not a necessary requirement for the perpetuation of a more pure form of capitalism, the existence of concentrations of high unemployment and direct economic distress are likely associated factors (Clark, 1980). For localities involved in the resulting politics of local economic development, no matter where they are located along the urban-rural continuum, the implications are equally noteworthy. Cox and Mair (1988) have found, for example, that restructuring has left a heightened interest in competition among localities, the evolution of local business coalitions, and a coalescing of local community factions around business coalition strategies. One of the critical results has been that the intensity and patterning of metropolitan development is guided increasingly by local processes (Wilson, 1991).

In dealing with quality of life aspects we need to consider this at several levels, following the example set by Maslow's "hierarchy of needs". On the lowest level we identify the basic needs for survival, considered to be about the same for all individuals. Examples of qualities for survival are food, clothing and water (note Survival Level in Figure 1.1). On the next level we have needs for safety, such as income, housing, energy, health care and welfare services. On the third level capacities for social contacts are stressed. Here, of special importance, is accessibility to education and resources for transportation and communication. The fourth level may be labelled self reliance, which can be promoted via travelling and participation in production and/or consumption of culture, but also through self employment. On the fifth, and the highest level, we have self actualisation. This means freedom to develop own ideas, and take individual initiatives to realise them.

In Sweden, the welfare state model has provided the inhabitants of ISERs and other regions with a relatively good quality of life on all levels (or at least on the four lowest). Focus has been placed on developing collective solutions suitable for families, especially those with children, and for the elderly. Policy and planning strategies have emphasized service needs for survival, safety and social contact. Compared with the core areas of Sweden, ISERs today obviously are lagging behind in higher education facilities, in variety of job options on the local labour market, in rapid transportation and communication modes, and in cultural amenity options. However, in many respects some medium sized cities are facing similar constraints.

Until the early 1980s, differences in living standards between labourers and civil servants was decreasing. This was a result of a combination of equalisation efforts in wage agreements and increased public transfers. The large expansion in the number of civil servant jobs in the public and private sector since the 1960s has also meant a stronger gain with respect to social class than in other Western European countries. Employment increases during the 1970s and 1980s were reflected mostly by female employment, while the employment rate among men decreased slightly.

Compared with other countries household incomes in Sweden are more evenly distributed. Estimates are that more than 50 per cent of the variation in income is eliminated by public intervention, while the corresponding figure is less than 40 per cent in most other countries (SOU, 1992:19). Until 1980, incomes increased in fixed prices. After that there has appeared a widening gap between the poorest and the richest. Another pattern is that across all socioeconomic groups and in private as well as public sectors, men experienced larger income growth than women and wage differences between men and women with the same level of education and work experience increased.

In spite of stagnated income growth during the 1980s, living conditions improved. For example housing, car ownership, holiday trips and technical equipment in the household increased. Workmen's dwellings approached the standard of civil servants dwellings. The share of households living in single family houses increased.

Living standards among the youngest have been in relative stagnation since the 1970s as a consequence of longer education, related higher debts for post-secondary school education and increasing unemployment problems.

In general, the oldest generation has experienced an increase in living standards, although pensions are particularly low for those who have contributed little from working life. This is also reflected in consumption patterns. Among most retirees we can observe a high level of consumption including mobility by cars and other transport facilities between a great variety of activities.

A recent national study (Socialstyrelsen, 1994) provides us with a general overview of changing living conditions. Surveys of living conditions reveal that the number of "good" jobs (i e jobs with a meaningful content and which can largely be managed by the employee) actually increased in the 1980s due to structural change from jobs in the goods-handling industries to jobs in the service sector. However, this development did not impact the female labour force. The 1980s appears have led to a polarisation: the number of jobs with positive characteristics increased, but so too did the number of jobs with negative characteristics.

The rate of early retirement among immigrants increased, especially among women from countries in southern Europe. This is explained in part by the fact that immigrants are over-represented in those professions where early retirement is more common. Among native Swedes, the proportion of early retirement remained the same through the decade.

The economic crisis of early 1990s has caused an extremely high unemployment severely hurting both men and women. Demand for labour decreased in most sectors and unemployment increased. Unemployment is highest among young people and immigrants and is also prevalent among single parents. In households with two adults it is once again becoming less common that both are economically active. For young couples and immigrant families who entered Sweden in the 1980s, it is becoming more common that only one is employed. Long-term unemployment is increasing and chances of finding a regular job decrease as the unemployment period increases. Immigrants are less likely to enter the regular labour market and also less likely to have access to labour market policy measures. The proportion of economically active handicapped people is also decreasing. The number of unemployed without unemployment insurance is increasing and half of the foreign citizens and one third of the young are not covered by unemployment insurance.

In addition to open unemployment, there are increasing numbers of persons with weak links to the labour market - under-employed and those in labour market programs. In total, these groups by far outnumber the number of unemployed; even the number of unemployed plus the number of early retired. For example, in 1993, 400 000 persons had temporary and insecure jobs, compared to the open unemployment of 350 000 persons. These groups are at risk of unemployment or of becoming permanently weak and marginal employees.

Age, gender and class also influence income. Thus, low incomes are numerous among young people, but also among the very oldest with many close to the poverty level. Disposable incomes below poverty level are more common among blue collar workers and although more men than women are below the poverty level, larger numbers of women are closer to the border of being classified as poor than men. Poverty is more common among foreign citizens, especially from countries outside Europe, than among Swedes.

Income transfers reduce the impact of poverty, first of all among the elderly but to some extent also among the young. The share of poor single parents was reduced in the 1980s. As much as 80 per cent of single parent families are moved out of poverty by social transfers (i e regular programs, not social benefits).

The proportion of clients receiving social benefits remained stable throughout the whole post-war period. This changed during the economic boom of the 1980s, however, as the number of households receiving benefits actually increased. As a consequence, costs increased due almost entirely to increased payments to immigrant households. This was partly a reflection of increased refugee immigration and the formal restriction that refugees waiting for asylum are not given working permits.

Only a minority rely on social benefits for long periods; most often families need temporary economic support. During the 1980s, the proportion of elderly people receiving social benefits was reduced, while the proportion of immigrants increased. Until the mid 1970s, elderly people constituted more than 15 per cent of those receiving social benefits. Currently the corresponding figure is 7 per cent.

Shortage of cash and difficulties paying for food, rent, etc., are more common among blue collar workers than among white collar workers and more common among single parents and immigrants.[2] In the 1980s, single parents especially experienced decreasing economic standards. The situation for immigrants, however, remained largely the same throughout the decade. Physical living conditions for children, in general, degraded in the 1980s; this is true for children in families within all socioeconomic groups and ages. The gap between children with a single parent and those with two parents increased in this sense.

As many as 500 000 children (more than 25 per cent) are living in families with reduced economic resources. Almost half of these children in the 0-15 year age group were in families whose disposable income falls short of the calculated living costs for that household type - for every fifth child with a single parent the disposable income of the family is below this level. However, from the mid 1970s to 1991, the proportion of families below the poverty level (below the norm for social benefits) was reduced from 21 per cent to 9 per cent.

---

[2] For example, immigrants are twice as likely to be unable to raise 12 000 SEK cash than Swedes are (Socialstyrelsen, 1994, p. 76).

In the 1990s, the situation for single parents has further deteriorated. Half of all single parents lack any cash reserves and almost as many have difficulties paying current bills. In the 1990s, the costs for social benefits increased as well as the number of clients receiving benefits. The number of native Swedes receiving benefits is now also on the increase.

According to most indicators, housing standards in Sweden are good and most people live in modern and commodious flats almost all with housing contracts. Thus, most "classic" problems associated with housing have been eliminated. In the 1980s, overcrowding continued to decrease and class-related differences were reduced in terms of housing quality. In the late part of the decade, however, young people faced difficulties entering the housing market, especially in metropolitan regions.

Housing segregation, both socioeconomic and ethnic, increased during the 1980s, mainly in metropolitan regions. The number of "mixed" housing districts decreased after 1985. An increasing number of communities in the periphery of metropolitan regions have an overrepresentation of immigrants and groups with meagre general resources. Many of these areas are deficient in their built environment and provide poor services and maintenance. People living in such areas report dislike and even fear regarding their environment.

The recession during the early 1990s, with high unemployment and increased economic pressure on many households will probably contribute to reducing housing standards for some families. It is necessary to track the changing conditions for those who are already living in overcrowded flats, i e families with children, immigrants and groups with limited material resources.

Deregulation of the housing market, reduced subsidies and market-orientation of rents is already influencing families as they attempt to maintain their housing standards. Young people, families with children and single parents are especially sensitive to changes in incomes and costs. On average 20 per cent of disposable income is used for housing costs in Sweden, more than in most countries in Europe. There is evidence that recent immigrants and young people now face increasing difficulties entering the housing market. The homeless are also becoming more visible with one estimate that they now exceed 10 000 in the country as a whole.

Housing grants are given mainly to families with children. For single parents and on average, 40 per cent of housing costs are covered by such grants. Since demand for such grants are increasing while public resources remain the same, it is likely that many of the families receiving grants today will see a reduction within the next few years.

In summary, housing segregation is expected to increase as a consequence of high unemployment, reduced economic resources, refugee immigration etc. It is likely that differences between attractive and less attractive housing areas will increase further in metropolitan regions. The classic problems concerning housing, overcrowding and low standards have been replaced by problems concerning the housing environment.

As a general conclusion, state and local government has taken a much wider range of responsibility for the quality of life in Sweden than in the US. As a consequence of the market model practised in the United States, socio-cultural conditions have evolved as a system tied much more closely to the centralising tendencies of relatively unfettered private enterprise. One of the comparative differences has been the relatively greater wealth accumulated in metropolitan growth regions.

2 PROCESSES RESHAPING THE SPATIAL STRUCTURE

2.1 Introduction

The purpose of this chapter is to explain and synthesize the on-going processes that work towards a restructuring of the prevailing spatial structure. The basic thesis is that the result of the interplay between these processes leads to the emergence of intermediate regions, with specific characteristics distinct from both the central and the more peripheral regions. Most future-oriented regional studies seem to focus on either core or peripheral regions. Therefore, it is a challenging task for us to argue that in many developed countries, regions between highly urbanized regions at the one hand side and peripheral regions at the other, are facing a new situation in the 1990s. This is due to a number of factors; general economic development, localisation shifts of private enterprise, new household preferences, new technology and the impact of public policy. These factors induce changes in the function of such regions, partly as a consequence of new actors making demands on the territory. Functions are described in terms of goods and services production, residential and recreational assets.

2.2 From a Resource-Based to a Network-Based Economy

Fisher (1933, 1939) was one of the first to formulate a theoretical framework on general development trends. He posits three development stages to describe the transformation of societies. The initial stage has a dominant natural resource sector. In the second stage the industrial sector dominates, accompanied by accumulated knowledge and prosperity. In the third stage service production becomes the leading employment sector.

Bunte et al (1982) have described the long-term development of a sparsely populated municipality typical of the Nordic countries (Figure 2.1). The structural changes follow Fisher's general transformation schema. Especially worth noting is the rapid transition that has taken place since 1950. One of the consequences of this process is that the fringe areas are losing their uniqueness based on the exploitation of available natural resources. Expanding activities - industries and services (especially welfare-motivated services) - develop their character in a way similar to those corresponding activities which are located in areas with more central geographical positions.

In the Nordic countries in general, the period since the 1960s has been characterized by a considerable shift of new functions to the public sector and a marked improvement of educational options and health and social services. This process has restructured the entire labour market. Women have entered the labour market on conditions similar to those of men. In a large proportion of families both the man and the woman have a paid job. Many new services have replaced previously unpaid jobs performed by housewives at home. Consequently, the variety of job opportunities within daily commuting distance and access to local

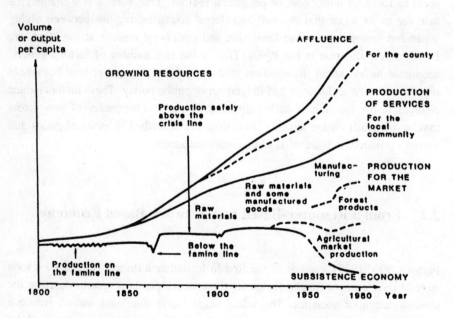

Fig. 2.1. The transition from a subsistence economy to a modern three-sector economy in a Swedish sparsely populated municipality. Source: Bunte, Gaunitz and Borgegård (1982).

day care services, schools and health services have become important factors guiding young families' decisions as to where to settle. At the same time, explicit and implicit regional policy has been to provide all municipalities with equal supply of services, a variety of private and public housing options and a diversified labour market. To a certain extent, this has reduced the incentive to migrate extensively. Partly as a consequence of this, the general outline of the Nordic settlement system has remained surprisingly intact during the process of industrialisation and deindustrialisation, The main change has been the local and regional rural-to-urban shift rather than any shift in the national urban system.

However, some authors stress the dynamics of the location pattern. Andersson (1986) argues that structural changes in production, location, trade, culture and institutions are accompanied by slow but steady changes in logistical networks. In this transformation process three components play a central role:

(i) emerging applications of information technology are implemented in production, control systems, administration and decision-making;
(ii) new telecommunication and datacommunication systems form efficient and rapid interactive links over both short and long distances;
(iii) the demand for fast transportation modes grows rapidly, focusing investments in transportation infrastructure on higher integration between local, regional and interregional transportation networks.

The concepts "information society" and "knowledge society" are used in the literature to stress important new links and node qualities in the network society. Production in different locations is linked together by volatile flows of commodities, people and information. Attractive sites for plants belonging to modern production systems offer an intensive and diversified local network of production linkages. They also provide opportunities for participation in national and international networks by having good access via high standard fast transportation and communication modes such as air transport and telecommunications. In this context we observe the trends listed in Table 2.1.

For many types of functional regions this development process represents a threat. According to Johansson (1989) "the on-going technology change implies a reorientation of economic activities towards knowledge-oriented production, advanced industrial, commercial and financial services, and systems design. Many small and large urban regions all over Europe are stigmatised by their mature specialisation in fields which are becoming out-dated and rigidities of infrastructure and human knowledge tend to prevent rejuvenation".

Table 2.1. Some general trends in the economy.

From:	To:
Resource oriented locational preferences	Environmental oriented locational preferences
Central places and local labour markets	Networks and complex commuting patterns
Rigid production and cheap labour	Flexible production and qualified labour
Labour-hunting firms	Place-hunting firms
Stable transportation and communication links	Flexible transportation and communication links
Slow transportation modes and storage	Rapid transportation modes and just-in-time delivery
Strong governmental intervention	Market orientation

Depending on their economic structure, some regions will adjust positively and others negatively to the new conditions. Major structural prerequisites for adjustments in either direction are listed in Table 2.2 (Hall, 1990; Johansson and Karlsson, 1990; Törnqvist, 1988).

We draw the following conclusions. First, there is no specific threshold marking the minimum requirements for a successful network economy; requirements are more of a qualitative than of a quantitative character. Successful results depend to some extent on local conditions and to some extent on competitiveness in widening international markets. Modernising production systems could threaten self-sustaining development in many ISERs as well as in urban regions where resource-based industries may become outdated. The needed replacement to maintain population (which is considered a primary goal for practically every municipality in Sweden) and to guarantee acceptable living conditions requires strategies and places stress on qualities different from those which formed the requisite conditions for the declining production base. Local labour markets and local governments must find their specific new role. To mark out a successful future pathway is a challenge. Castels (1989, p. 351) formulates the problem in the following way:

> "This is possibly the most difficult dimension to integrate into a new strategy of place-based social control, since a precise and major characteristic of the new economy is its functional articulation in the space of flows. Local governments attempting to restore social control of the development process need to establish their own networks of information, decision-making, and strategic alliances, in order to match the mobility of power-holding organisations".

Table 2.2. Key factors in regional adjustment processes.

Negative adjustment:	Positive adjustment:
Export specialisation	Innovative import activities, higher-level service base
High share of price competition firms	High share of dynamic competition firms
High share of intermediate level institutions	High level of educational options
Few and stable linkages to other nodes	Complex and flexible linkage pattern: high quality transport node for people (air transport, high speed rail system), well developed telecommunications infrastructure
	An urban nucleus characterized by plurality of activities and meeting-places
	Attractive residential options appealing to highly skilled labour

Table 2.3. Infrastructure variables with significant impact on regional development. Source: Revision of Johansson and Karlsson (1991).

Local and intraregional infrastructure resources:	Regional and interregional networks:
Education and occupation categories of the population - with an accessibility characterisation	Accessibility in the road system to international markets
	Telecommunication capacity
The built environment as a system of premises, service centres and housing areas	Postal distribution capacity
	Frequency and speed of interregional trains
Accessibility networks - flow capacity and quality of road networks - public transport capacity - accessibility between centres/ localities within the region	Airport capacity and accessibility to airport
	Accessibility to other region's airports

Assuming an on-going tendency towards an increasing share of footloose firms, the future structural stability of localities and regions will depend very much on quality of infrastructure, defined in a wide sense. Areas where low levels of investments in infrastructure occur will face severe negative consequences. Table 2.3 presents infrastructure variables which impact on the economic performance in regions and act to attract place-hunters.

2.3 Quality of Life Improvements

As indicated in the discussion on different levels of service needs, collectivism versus individualism and public versus private organisational modes are important factors when comparing efforts to provide improvements in quality of life in

Sweden and in the US.[1] Table 2.4 illustrates possible approaches to promoting quality of life.

The traditional Swedish welfare model emphasizes the combination of collectivism and public institutions. In the on-going reorientation, privatisation, decentralisation and deregulation are all used. The welfare services developed in the ISERs are heavily dependent on transfer of governmental financial resources, which indicates low interest for private risk-taking projects. Currently, our only knowledge is that financial support from the government will be reduced substantially during the next decade. In light of this, one possible and attractive alternative for ISERs is a combination of decentralisation and deregulation which promotes higher adjustment to local needs and goals.

A third aspect of the quality of life is the combination of the two dimensions collectivism versus individualism and materialism versus postmaterialism (Table 2.5). These combinations stress societal values among different generations and categories of people. Though we expect that large differences may be found both within ISERs and other portions of the spatially defined continuum, we are able to make some generalisations. A substantial proportion of the population in the Swedish ISERs appear to have a life style focused on materialism and collectivism. Traditionally a great variety of more or less social associations are present in most of the local communities. However, among the youngest generation there is a movement toward increased individualism and strengthened emphasis on materialism. Traditional associations are facing recruitment problems. As yet, there has not been as strong a diffusion of postmaterialistic ideas and values, for example in the form of concern for environmental qualities, as in urban areas.

Table 2.4. Divergent approaches to promoting quality of life improvements.

Promotion of:	Organisational mode: Public	Private
Collectivism	Welfare state	Privatisation
Individualism	Decentralisation and deregulation	Privatisation and deregulation

[1] This section is largely abstracted from Gade, Persson and Wiberg (1992). See also Gade (1991).

Table 2.5. Quality of life directions of change.

	Collectivism	Individualism
Materialism	Emphasis on basic needs	Emphasis on materialistic satisfaction
Postmaterialism	Emphasis on environmental, ethical and moral issues	Emphasis on personal freedom

For the social welfare states the German social philosopher Habermas believes that it is time to think about those conflicts that deviate from the traditional welfare state pattern of institutionalised conflicts over distribution. He discusses the shift in values and attitudes characterized by Inglehart (1979) as the "silent revolution". Here the "old politics" of economic and social security, and internal and external military security, has become the "new politics" of quality of life, equal rights, individual self-realisation, participation and human rights. Adherents of the "old politics" are apt to be concentrated among trades and business people, and lower- and middle-class workers, as well as farmers. On the other hand, those who are more inclined toward the "new politics" are apt to be found in the new middle class, among the better educated and younger generation. This suggests to us that the world of "old politics" is apt to dominate in more rural areas, while the "new politics" finds greater support in urban areas. In other words, the contrasts in these perspectives are operant along a spatially defined traditional-modernity continuum that is coexisting with the urban-rural socio-economic gradation in quality of life conditions.

As evidence of a changing social order, some authors have pointed to a dramatic decline in class-related voting in Scandinavia (Worre, 1982; Holmberg and Gilliam, 1989). Others have emphasized a drop in group consciousness. Korpi (1978) found that younger Swedes had highly selfish motivations for their union membership whereas older workers saw union membership in class terms. In the early 1990s we may observe a decline of solidarity in the union movement in Sweden, and with it a further disintegration of community values and identity.

For Great Britain, Goldthorpe et al (1968) found the suggestion of a movement away from participation in sociable communities toward a more privatised form of social existence, in which the economic advancement of the individual and family becomes of greater importance than membership in a closely knit local community (1968; see also the discussion in Nielsen, 1990). In so far as this is valid throughout Great Britain, one may consider this to be further evidence of the integration of urban-rural values, as exist to an even greater extent in the United States.

Likely due to the readily observable social class characteristics of the inner ISERs (most adjacent to metropolitan areas) in the United States, there is found the clearest expression of postmateralism in the country. With earlier materialistic objectives well in hand, people are moving increasingly toward achieving the "new politics" of quality of life, individual self-realisation, participation and human rights. These comparative shifts in values and attitudes are found increasingly among the better educated and younger professionals who comprise a significant proportion of the population in the ISERs.

We suggest that increasing individualism and postmaterialistic values will stimulate the power of attraction of certain parts of the ISERs. These areas provide manifold unique combinations of urban and rural resources that are available for individuals and households choosing and experiencing different lifestyles. In a variety of aspects these regions also have a standard similar to urban regions. In general, the housing qualities - compared to housing costs - is more favourable in ISERs, mainly because of the relatively good supply of single family homes in attractive environments. This makes it relevant to talk about a greater competition between ISERs and urban centres. The competitive strength of the ISERs is founded on those positive rural qualities which they can develop and maintain.

Following Höjrup's (1983) classification, the life styles of the wage-earners, the career-oriented persons, and the self-employed living in ISERs differ in their way of making use of, and appreciating, available urban and rural qualities. In this way the quality of life in these regions may differ not only among regional contexts but also among individuals. Possible profiles of some typical lifestyles are outlined in Table 2.6. Here the intent is to show the extent to which individuals in each of the three lifestyles perceive the utility of urban and/or rural qualities in housing, income generation, consumption of services, social life, and recreation.

Table 2.6. Typical lifestyles and utility of urban and rural qualities.

	Wage earner	Carrier oriented	Self employed
Housing	U+R	U+R	U+R
Income generation	U	U	U+R
Consumption	U	U	U
Social life	U	U	U+R
Recreation	U+R	U+R	U+R

U=urban qualities, R=rural qualities

2.4 Emerging Spatial Structures

In a discussion of long-term structural urban trends, Hall (1990) identifies four interacting processes appearing in Western Europe as well as most other countries in the industrialized part of the world.

(1) In a local context there is a process of deconcentration of jobs and people from city cores to suburbs.

(2) Within the economy there appears a deindustrialisation in the form of employment reduction in resource-based and goods-handling industries and expansion of information handling and knowledge-based industries, involving a shift from industries with a strong link to the local and regional environment to footloose industries.

(3) There is a reconcentration from large metropolitan cities to medium-sized city regions. According to Hall and Cheshire (1988) many of these fast growing medium-sized cities are acting as employment magnets, while population increases take place in the hinterlands with a rural appearance but not remote from the urban cores.

(4) Finally it is possible to observe new growth corridors between city cores with intermediate rural areas as attractive and well integrated parts of the corridor. Often an investment in transport infrastructure has created a motive for location of activities and increase of settlements in these corridors. In some cases a new airport has played an important role, both real and symbolic, as initiator of a positive development. In other cases,

improved standards of roads facilitating commuting has played a similar role. A third alternative infrastructure strategy has been to link city cores with a high-speed rail system.

These trends, together with those previously discussed, allow us to draw the same conclusion as Hall (ibid., p. 181). Many economic activities no longer need traditional cities, developed according to earlier production and logistical principles. "People want to get out of them, or perhaps have to get out of them. The people and the jobs come together in smaller places arrayed in belts of growth, typically in places little touched by earlier industrial revolutions."

In a concluding section of the book *Cities of the 21st Century* (Brotchie et al, 1991), is discussed how the new information technology is leading to the emergence of complex substitutions in various fields and changing the character of goods, markets, work environment and job location. The shift to information-based industries is giving individuals greater freedom in locational choice. Working arrangements as well as spatial organisation of activities can be more informal and flexible. Much greater concern can be placed on individual preferences for different urban, rural and recreational amenities.

Throughout Europe we find a considerable number of examples of entrepreneurs and founders of new firms migrating out of big cities into more favoured regions; those with better climate, scenery, and/or other quality of life advantages. This urban-rural shift is marked within high technology firms, which often are serving specialised global markets. According to Keeble (1991) "there is a clear, consistent and striking gradient in recent high technology employment change between urban and rural counties of Britain". Between 1981 and 1989, the most densely populated conurbations in Britain lost 16 per cent of jobs in high-technology industry while less urbanized counties gained 7 per cent and rural counties gained 12 per cent of the new jobs (ibid.).

As a consequence of more informal organisation of work the daily activity pattern among people is becoming more flexible. The most obvious substitution potential in a short-term perspective is replacement or reduction of daily commuting between home and office for white collar workers. The dominant pattern of a uniform office tradition, a result of the industrialist era, is facing organisational and psychological barriers. Both among employers and employees, adjustment must take place in several different ways. The need to achieve substitution effects, induced from pressure to increase competitive strength, may have considerable consequences for individuals in terms of organisation of daily life. New types of co-ordination patterns between employer and employees and between and among colleagues will also appear.

The average speed of transition in Swedish ISERs in terms of in-migration share of population is approximately 20 per cent over a 5-year period. Today the relative proportion of white collar workers with flexible job options living in rural areas and small towns is, in most cases, marginal. Attraction of more white collar people to ISERs is a matter of provision and maintenance of high environmental qualities, local basic services for frequent consumption, good standard of roads to nearby centres and good qualities of telecommunications.

2.5 Function and Quality of the ISER in Sweden

We suggest a definition of the ISER (intermediate socio-economic region) which is based more on its function and qualities than on other aspects such as size, density and growth rate. The ISERs represent the linkage zones between active and ever changing metropolitan regions, and potentially stagnating peripheral regions, relatively devoid of socio-economic growth and change. For Sweden the top down national development planning approach has insured inclusion of the intermediate regions in functional labour markets. This is a type of area where there is an acceptable population base for most types of daily demanded services and an ordered (even if limited) wage labour market, although at the same time conditions are poor for advanced services to households and businesses.

We want to emphasize that our interest in the ISER is not identical, conceptually, to earlier expressed interests by regional planners in the potential role of medium-sized cities in the development process (Hansen, 1970; Roepke and Freudenberg, 1981; Wiberg, 1992b).

We have chosen an explorative research approach to the ISER concept, rather than a normative approach. Hence we have used several cluster analyses in Chaps. 4 and 5 in order to identify homogeneous regions in terms of different relevant characteristics. By changing the set of variables describing each cluster, we reach a more multifaceted description of central, intermediate and peripheral regions. By using different clustering procedures and variables we try to neutralize some of the subjectivity which is involved in every cluster analysis.

3 PATTERNS OF SECTORAL AND SPATIAL CHANGE

3.1 Internationalisation and Locality

Current economic processes include the globalization of trade and production networks. This is a trend just as important in North America as in Europe. Economic actors in large as well as in many small corporate firms operate in geographically widespread markets, looking for optimal locations according to production costs and marketing. This, however, does not necessarily mean a concentration of their activities. Just as likely is a pattern of deconcentration. Information on conditions in different sites is readily accessible through both formal and informal channels. The process of relocation of economic activities is facilitated by the rapid structural transformation from goods-handling to service

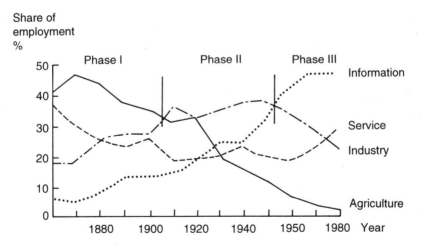

Fig. 3.1. Development of employment structure on the labour market in USA 1860-1980. Source: Piatier, 1981.

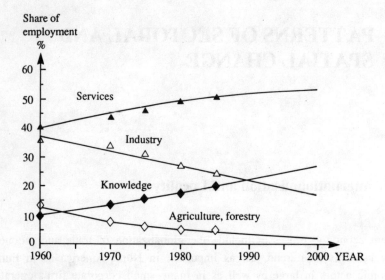
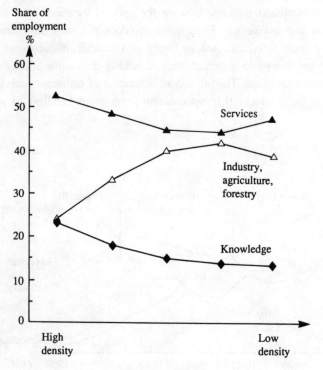

Fig. 3.2. Development of employment structure on the labour market in Sweden 1960-2000. Source: Larsson, 1992.

branches. The importance of resource-based production is replaced by production of goods and services with a substantial knowledge content (Figure 3.1 and 3.2).

The structure of manufacturing industry in various types of labour market areas in Sweden is presented in table 3.1. The frequency of sheltered, labour, capital, knowledge and R&D-intensive industry is shown for four types of regions. The extremes are metropolitan Stockholm with intensive knowledge- and R&D-based industry and peripheral regions which are largely based on labour-intensive production.

Private firms seeking optimal new locations and places to expand look as much for qualified labour and rapid communication networks as for access to physical inputs. This means that the location has to supply a wider array of inputs than is usually available and accessible in either metropolitan or peripheral locations. Evidently, some of the intermediate regions can provide such an array of material and non-material facilities. In spite of a global market, local, or microregional, conditions are becoming more important.

Table 3.1. Structure of manufacturing industry[1] in various types of local labour markets in Sweden 1990. Per cent of employment in manufacturing industry. Source: Statistics Sweden.

Spatial domain	Intensity on:				Sheltered	Unclassified	Total
	R&D	Knowledge	Capital	Labour			
Stockholm	30	18	5	11	30	6	100
Medium sized cities	9	25	16	25	22	3	100
Intermediate regions	6	30	11	30	21	2	100
Periphery	3	14	1	50	29	3	100

1 Categories of manufacturing firms according to a classification by Ohlsson and Vinell (1987). R&D-intensive industries have a high per cent of employees with a formal scientific education and comparatively high R&D costs. Knowledge-intensive includes industries with a high per cent of engineers and other qualified technicians. Capital-intensive has comparatively high capital input, often heavy traditional industry. Labour-intensive employs a relatively large number of unskilled workers. Sheltered includes the parts of food, wood and building material industry which have small exports (Ds 1992:81).

The internationalisation process sometimes means that many decisions at the firm or corporate level are made within the framework of joint ventures. Companies and actors representing different cultures and different institutional backgrounds compromise in these decisions. Thus, more location decisions are being made from a multidimensional and even multicultural viewpoint. Sites that exhibit several attractive conditions for production and residence are probably more successful in attracting international ventures. Environmental, cultural, natural, aesthetic and security factors are more important than pure economic factors.

The economy of Sweden is, to a large extent, dependent on large flows of primarily goods but also services across its national borders. This dependence, which can be measured as an export share of production value as well as an import share of domestic consumption, has almost doubled since early 1970s. However, since 1983 there has been rather stable export and import shares (Figure 3.3).

In terms of foreign investments in Sweden and Swedish direct investments abroad, there was a dramatic increase during the 1980s. Between 1980 and 1990,

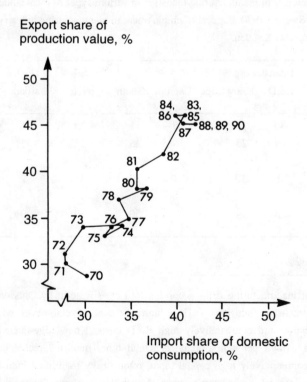

Fig. 3.3.   Swedish dependence on export and import. Source: The National Accounts.

foreign investments in Sweden increased from USD 400 million to USD 2 billion and investments abroad increased from USD 1.2 billion to USD 14 billion. After a considerable peak of foreign investments in Sweden in 1991 (USD 6.4 billion), we have during 1992 and 1993 experienced a drop of both types of investments back to approximately the same level as in early 1980s. The economic recession is one important explanation both for this and for the decrease in Swedish investments abroad.

Table 3.2 provides information on the relative spatial distribution of foreign investment in recent years. Västerbotten is a county in northern Sweden with a strong character of modern service sectors in the central zone, of resource based industries in the other zones. Västernorrland is also located in northern Sweden, but more dependent on traditional industry. More than the other counties, Uppsala county in central Sweden is characterized by knowledge-intensive sectors. Among the selected counties foreign acquisitions of manufacturing firms appeared with comparable relative frequencies in central and intermediate zones, but with significantly lower level in peripheral zones.

Since 1970, the sheltered public service sector has increased considerably in absolute and relative terms. This means that although a decreasing part of the Swedish economy has been involved in international competition, up to late 1980s this sector has been able to increase export share of production value. In the early 1990s, international depression in combination with structural changes caused severe problems for the dominant Swedish export sectors such as forest-based industries and car manufacturing. In the context of the international division of labour, there was widespread belief that Sweden would face problems in com-

Table 3.2. Share of foreign acquisitions of manufacturing firms in three Swedish counties 1985-92. Source: Aktuellt om fusioner, Swedish Competition Authority.

County	Share of foreign acquisitions, related to number of enterprises and employed()					
	Central zone		Intermediate		Peripheral	
	%	%	%	%	%	%
Västerbotten	1	(0)	4	(8)	1	(8)
Västernorrland	3	(20)	3	(2)	0	(0)
Uppsala	4	(6)	3	(2)	0	(0)

peting successfully in manufacturing associated with mature products and strong international price competition. After the liberation of the exchange rate of the Swedish "krona" (SEK) in 1992, some of the problems concerning high wage costs in Sweden were reduced. The consequences for exports from the traditional labour-intensive manufacturing industries have become obvious. Niches for expanding Swedish industries are however to be found in technology and knowledge intensive manufacturing and services.

Concerning the implication of technological processes in the reshaping of production landscape, development of transport and communication infrastructure technology over recent decades has opened a variety of new possibilities for the establishment of rapid and efficient contact and exchange relations. The US and Sweden are among the countries with the most developed telecommunications. Besides a high level of spatial diffusion of telephone lines and equipment the cost levels for telecommunications use are favourable for households as well as firms, in an international perspective. The most favourable conditions are found in Sweden where a low cost level is combined with a tariff structure where operating costs per unit time are the same independent of distance further than 90 kilometres within the country. Figure 3.4 presents a comparison of international telecommunication costs for firms in the OECD countries with fixed costs and operating costs shown separately.

The regional consequences of the new technological qualities are a function of priority given to the process of upgrading and how new options are distributed geographically. One strategy is to promote the exploitation of new potentials and the creation of new regional structures. Another strategy is to focus on existing bottle-necks and missing links in current structures. In reality examples of both strategies can be found in Sweden as well as in the US. Both types of strategies have been important in narrowing accessibility standards among extreme levels of human settlements.

The rapid technological development of information technology with its increased linking capacity and quality of telecommunications has created options for information-handling industries to locate more freely in geographical space yet still maintain competitive positions. Unfavourable transaction costs may be balanced by favourable locational costs. The potential for creating new regional structures using information technology is large but is facing resistance to change within existing organisations.

Modernising production systems could threaten self-sustaining development in many intermediate regions as well as in urban regions where resource based industries may become outdated. To maintain population and living conditions requires strategies and stress on qualities different from those which formed the suitable conditions for the now declining generation of production.

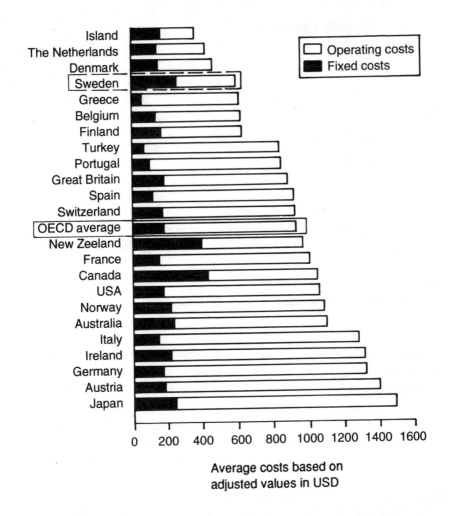

Fig. 3.4. An international comparison of telephone costs for firms in 1989. Source: OECD Tariff Comparison Model/ OECD: Performance Indicators for Public Telecommunications Operators, 1990.

The urbanizing process in rural areas is, to a large extent, a consequence of increased mobility of resources available to households and individuals. For example, the share of people who are commuting has increased dramatically during recent decades and, in many Swedish and US intermediate regions, is now a dominating pattern. Figure 3.5 illustrates how the average distance travelled by ground transports per day in Sweden has grown during the last hundred years.

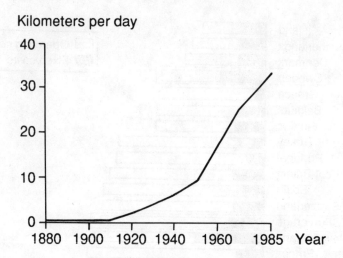

Fig. 3.5. Length of ground transport in kilometres per inhabitant and day in Sweden. Source: The Population, National Atlas of Sweden (1991).

Compared to Sweden, the US has been a society with high factor mobility both for a longer period and to a much greater level. For example, North Carolina has consistently elected governors, who emphasized road construction in their political platforms. Since the 1930s, heavy investments in road development projects have occurred, even in more sparsely populated areas. In addition, the evolution of a virtual megalopolis in the state's urban Piedmont region is, in large measure, due to the past twenty years of interstate freeway development. There has also been a much stronger promotion of private car ownership, compared to Sweden. One significant result has been the evolution of a metropolitan labour shed in North Carolina that is much more tied to individual commuting behaviour. In Sweden the central government has, via subsidies and sectoral investments, promoted collectivistic (in most cases public) transport solutions, and simultaneously, via taxes and regulations, slowed the increase in private car use. This significant difference in mode of access is undoubtedly one of the factors that helps to explain some of the differences found in the locational characteristics of intermediate regions in Sweden and the United States.

Against this background the intermediate regions may be facing different possible futures, ranging from marginalization to competitive development based on the unique combination of rural and urban resources and the specific living conditions that they provide for individuals and households maintaining or choosing different life styles.

3.2 Social Change and Preferences

We also must address the social processes involved in the restructuring of the centre - periphery hierarchy. The mobility pattern of the population is changing, partly as a result of the increasing two-person income dependence and professional carrier of the individuals. In large and congested urban regions as well as in small towns, there are limited possibilities for providing a close geographical connection between the place of residence and the place of work for more than one of the income-earners in the household. This has meant that intensive daily commuting is now a basic feature in both these types of regions. Locations which can provide less time-consuming commuting may prove to become more and more attractive for households and families looking for more and useful leisure time. Intermediate regions often can provide such a setting.

As a consequence of increased private income, more flexible working conditions and changing values, leisure behaviour becomes more geographically widespread and individually differentiated and qualified. The importance of access to different and spatially demanding leisure activities means that more households look for residential and work-place locations that allow for this behaviour.

The more flexible life-styles preferred by many individuals is also expressed in a more flexible family structure. This means that long-lasting relations are becoming less typical. This is related to a general shift in preferences which is reported from many western economies.

We can provide an example of variety in attractivity of different types of localities in Sweden in terms of migration behaviour. By examining the last fifteen years, which includes a period of stagnation in the Swedish economy around 1980, we get a reference to the discussion later in the book on consequences of a higher level of adjustment to market forces and reduced financial resources to maintain high spatial equalisation principles within the public sector.

The municipality of Skellefteå, located along the coast of the Bothnian Gulf, can be characterized as a traditional single core location with a strongly dependent hinterland (population 76 000 in 1990). The analysis uses data for the 12 parishes which comprise the municipality. As the physical size of the municipality is large (6 800 km^2) and the dominant node has a rather centred location, there is a strong consistency between borders of the municipality and a demarcation of the local labour market around the municipal centre. Consequently a dominant part of job options for people living in all parts of the municipality are available within the municipality.

The attractiveness of different parishes within the municipality has been calculated by comparing the number of in-migrants to the population of the parish. Different levels of attractiveness are calculated by analysing the behaviour of three categories of migrants:

Locally attractive: people who have migrated between parishes within the municipality.

Regionally attractive: people who have migrated to the municipality from other parts of the region (county of Västerbotten).

Nationally attractive: people who have migrated to the municipality from other parts of Sweden.

Figure 3.6 illustrates the distribution when average distance to the city core of the municipality is taken into consideration. The total attractivity shows an expected distance decay pattern. However, a dominant part of this is a function of local migration behaviour. Among people in the categories of regional and national in-migrants we do not find such a distinct distance decay pattern.

There are considerable differences in population among the parishes - highest is 20 000 and lowest is 300. These differences in size have impact on flexibility to meet in-migration preferences in relation to out-migration and house construction

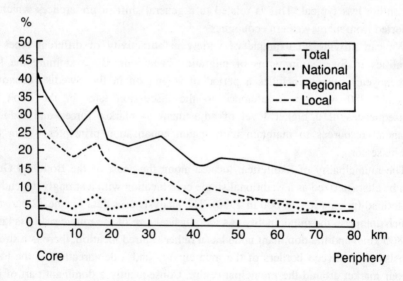

Fig. 3.6. Attractivity along the core - periphery axis in terms of in-migrants during 1986 - 90 related to population size. The municipality of Skellefteå, Sweden. Source: Statistics Sweden.

activities. In Figure 3.7 in-migration ratio (migrants during a five year period divided by population) is compared with out-migration ratio and population size.

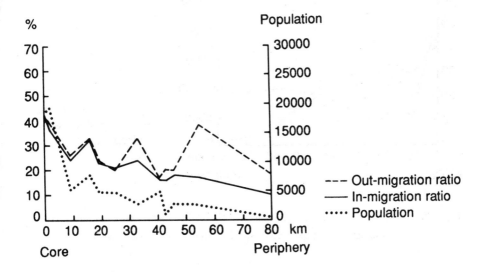

Fig. 3.7. In-migration and out-migration ratio 1986-90 along the core - periphery axis in the municipality of Skellefteå, Sweden. Source: Statistics Sweden.

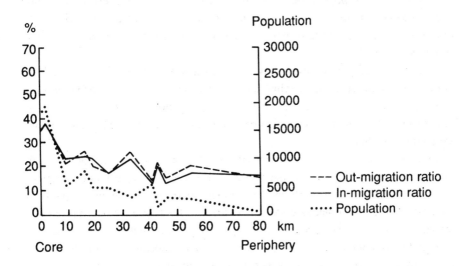

Fig. 3.8. In-migration and out-migration ratio 1981-85 along the core periphery axis in the municipality of Skellefteå, Sweden. Source: Statistics Sweden.

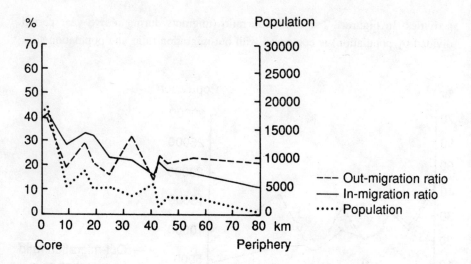

Fig. 3.9. In-migration and out-migration ratio 1976-80 along the core - periphery axis in the municipality of Skellefteå, Sweden. Source: Statistics Sweden.

In all parishes located further away than 25 kilometres from the municipal core the out-migration ratio during the latter part of the 1980s was higher than the corresponding in-migration ratio. In fact, since 1975, the peripheral parts have appeared as losers of population. The most balanced relation between in-migration and out-migration in all parts of the municipality appeared in the early 1980s during the latter part of the mentioned stagnation period in the Swedish economy (Figures 3.8 and 3.9).

From this example we may draw the conclusion that small scale localities during short term economic decline have been as competitive as urban centres in keeping balance between out-migration and in-migration. It is however important to observe that during this stagnation period the public sector acted as a strong and expanding backbone.

Another factor illustrating attractivity is the price level of single family dwellings. Figure 3.10 presents the average price level for single-family dwellings sold between 1983 and 1989 in the different parishes of the municipality of Skellefteå. A distance decay pattern is evident, reflecting locational preferences among in-migrants. The price level is up to five times higher in the core area. Within the intermediate region we also find a considerable variation in price levels but all average prices are lower than the average prices in the core area and higher than in the peripheral part of the municipality.

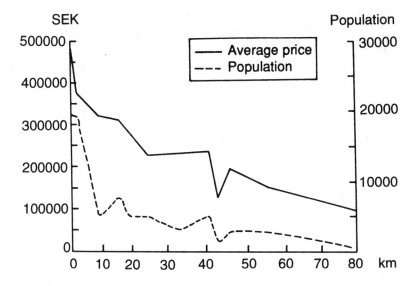

Fig. 3.10. Average price level for single family real estates along the core - periphery axis in the municipality of Skellefteå, Sweden. Source: Statistics Sweden.

An examination of how subjective living conditions - the quality of life - are experienced by families migrating to rural parts of intermediate regions, reveals a creative mixture of attraction and adjustment to local conditions. Ennefors' (1991) survey of 50 households shows that:

o   Some households have taken their decision to migrate to rural areas on the basis of a consideration of both rural and urban qualities. The primary driving force is the wish to settle down in a milieu with a rural style. However, the final decision to migrate is taken after having checked that resources of an urban character, such as food stores, a bank and a post office, a day care centre and schools are within an acceptable distance from the place of residence which is being considered.

o   The most important pull factor on people migrating to urbanized parts of rural areas is high landscape qualities.

o   Qualities related to the place of residence are often given priority over job opportunities and incomes for household members.

o   The rural resources which have been important factors of attraction in the past seldom play an influential role in maintaining the materialistic standard of the households. Activities such as cultivating, fishing and hunting are popular among many households but have a symbolic rather than an economic value. Substitutes are often readily available in local food stores. The main motive behind the active exploitation of rural resources is often to support the maintenance of family and community traditions and thereby also the strengthening of the household's own identity. For people who have previously experienced life in big cities, the rural milieu offers exclusive advantages such as cleaner air and water, seclusion and open landscape views.

o   Except for those households with children in remote locations who need public transportation to day care centres and schools, these in-migrating households do not demand much more infrastructure than good roads to higher level centres.

o   In many situations, and often due to commuting to work, people do not use the most nearby alternatives for shopping.

o   The households often prefer to develop their own solutions to cover their service needs. In some cases they use their own knowledge and resources. In other cases they obtain complementary help from their neighbours. Examples of these kinds of service areas are commuting trips by private car, repairs, house renovations and day care for children.

o   The non-institutional service solutions households develop on their own or together with their neighbours are often highly valued among themselves. These activities confirm their belief that a good lifestyle is possible in rural areas, even if it can not be regarded as an optimal alternative from the point of view of materialistic standards.

Finally, in discussing social processes, it is important to discuss long-term changes induced by generation shifts. Social research makes evident that preferences and behaviour are attached largely to specific generation patterns. The experiences of each generation have a long term influence on social behaviour. Thus, the current oldest generation in most western societies is rooted largely in an old rural structure and economy and their preferences largely reflect the basic material and cultural conditions of that time. On the other hand, the intermediate generation is associated more with an urbanized regional structure and reflects the basic ideas related to urban culture. The new generation, however, may in many cases have a different value system, reflecting the experiences of the declining social and physical environment of bigger urbanized areas. This experience may lead to a more favourable opinion on small and medium-sized cities and localities within their local labour markets lacking most of the negative factors associated with the metropolis. It is evident that, in this sense, different countries are in different phases.

3.3 Distribution of Income and Services

The total service sector (i e including services under private and public management) has increased in Sweden as well as in other industrial countries. In the US, services account for 75 per cent of the total labour market and the present level in Sweden, 70 per cent, was passed in the US ten years ago.

Sweden has a substantially larger public sector than most comparable countries in the western industrialized world. This is true whether size is measured in terms of public consumption, tax amount or the proportion of public employment at the labour market. However, this has not always been the case. As late as the early 1960s, the public sector in Sweden was of medium size in an international comparison. At that time, countries like the Netherlands, France, Western Germany and Great Britain had higher public expenditures per capita. But while expansion in other countries stagnated in the period of economic recession in the mid 1970s, growth in the public sector continued in Sweden.

In relative terms, the US figures only represent 50 per cent of Sweden's current public sector. Expansion and development of the public sector in Sweden has primarily been concentrated in social services and health care. While 92 per cent of employment in the health service sector is administrated by the public sector, the corresponding US figure is only 19 per cent, the rest being privately administrated. Within social services, differences are even more marked: 98 versus 11 per cent. However, it should be noted that these differences concern the form of production, the way of financing and the distribution of services and welfare among different socio-economic groups, not the relative size of the resources that are put in.

The public service sector in Sweden has expanded throughout the whole 20th century. Growth was most intensive during the 1960s and 1970s with stagnation starting in the early 1980s. Employment expanded 69 per cent during the 1970s and dropped to 10 per cent during the 1980s. More than 90 per cent of the employment increase took place within the municipalities and county councils. To finance this expansion, taxes in relation to GNP increased from 40.5 per cent in 1970 to 56.5 per cent in 1990 (Ministry of Labour, Ds 1993:78).

Current policy documents (such as the Medium Term Survey of the Swedish economy) do not expect any further growth, but rather the contrary. The policy debate is oriented largely towards means to stop a galloping increase of deficit in the state budget. There are disputes on general level of tax pressure, taxation profile and priorities in public spending. High levels of income taxes and dependence on public services are often considered to be obstacles to economic growth.

To illustrate how government expenditures are distributed along a centre-periphery continuum, a crossection of 25 municipalities along the Bothnian Gulf have been chosen from Stockholm to the border of Finland. The total distance is 1 100 kilometres. Six of these municipalities have fewer than 20 000 inhabitants. The populations range from 7 700 to 659 000. The highest expenditures per inhabitants are found in the metropolitan area - more than 100 per cent above the average level of the other municipalities. Beyond the metropolitan edge the five municipalities with the highest government expenditures per capita show a variation in population from 10 000 to 85 000 (Figure 3.11).

In the analysis different types of expenditures have been divided into the following five categories (SOU 1989:65):

Transfers: expenditures for pensions and various types of social insurance.
Reaction: expenditures for regional policy measures and to cover various problems such as employment problems within private firms, low local potential for taxation and unemployment.
Complex: expenditures for military forces, support to housing and energy production.
Service: expenditures for administration, social care, service to private firms, labour market training schemes and public transportation.
Competence: expenditures for education, research and culture.

The categories titled transfer, reaction and complex represent expenditures for activities focused on provision of a basic welfare and social security. Figure 3.12 presents the distribution of these three expenditures. As a consequence of welfare policy aimed at providing all inhabitants a basic living standard regardless of where they live, we find small variations along the continuum. On a per capita basis, peripheral and sparsely populated municipalities are as well provided with resources as municipalities with strong urban characteristics. In most municipalities the transfer category represents 50 per cent of the total expenditures.

The category labelled reaction has a pronounced regional policy profile with a spatial distribution reflecting increasing need for intervention moving towards the periphery.

Per capita expenditures within the third category, complex, evidence considerable spatial differences due to the nodal character of military bases and energy production.

A concluding general remark about these three categories is that most parts of these expenditures are either devoted to permanent basic needs among people or provided to handle short term problems in local labour markets.

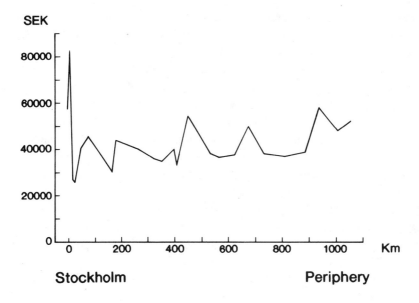

Fig. 3.11. Total public expenditures per capita along a centre-periphery continuum in 1985/86. Source: After SOU 1989:65.

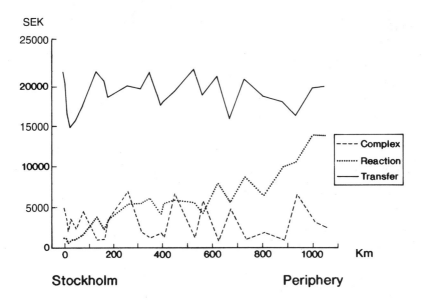

Fig. 3.12. Public expenditures per capita for the categories transfer, reaction and complex in 1985/86. Source: After SOU 1989:65.

Within the other two categories - service and competence - there appears a stronger future-oriented profile. There are obvious potentials to restructure local labour markets in higher education and labour market training schemes but also in quality and variety of culture and public transportation standards. The spatial pattern of governmental expenditures within both of these categories has a pronounced nodal pattern (Figure 3.13). Along the continuum we find the highest per capita expenditures in medium-sized cities which have leading roles as service centres for their regions. The consequences of the nodal concentration of resources to higher education is reflected in the rather low educational level of the labour force in municipalities beyond daily commuting distance to the university cities (see further discussion in chapter 4).

During the 1990s the transfer pattern has become more biased than during the 1980s. The regions which have received additional financial resources are mainly the metropolitan areas, major regional centres and the sparsely populated areas in the interior of northern Sweden (NUTEK, 1994). The distribution pattern give further evidence for our hypothesis of growing micro- and macroregional disparities.

As the pattern of change brings the Swedish socio-economic system into closer alignment with that of the United States it is certainly relevant to consider what

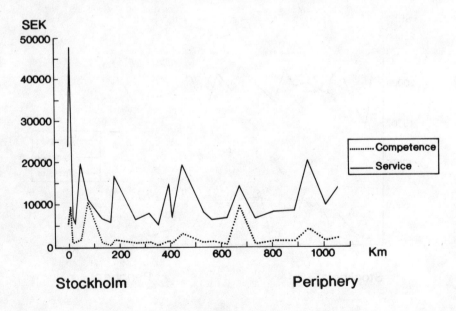

Fig. 3.13. Public expenditures per capita for the categories service and competence in 1985/86. Source: After SOU 1989:65.

has happened in recent years in the United States. As Lipset (1986, p. 452) has noted, "the American social structure and values foster the free market and competitive individualism, an orientation that is not congruent with class consciousness, support for socialist or social democratic parties, or a strong trade union movement".

Thus we see the current United States showing an increasing gap between wealth and poverty. Figure 3.14 provides an example of the widening gap at the state level. We are seeing American affluence shifting in its social condition toward upscale residential concentrations of business owners, corporate managers, skilled service industry professionals, and well-off retirees, and in its geographic condition toward those locations favoured by these types of individuals. Phillips (1990) builds a case for not just the re-emergence of the income gap, but for its pronounced developing identity between wealthy states/metropolitan regions and poorer areas, which are mostly rural, low income urban and rural peripheries and low income suburbs.

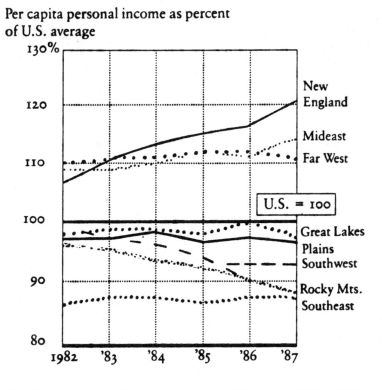

Fig. 3.14. US regional disparities widen during the 1980s. Source: Phillips, 1990, p. 187.

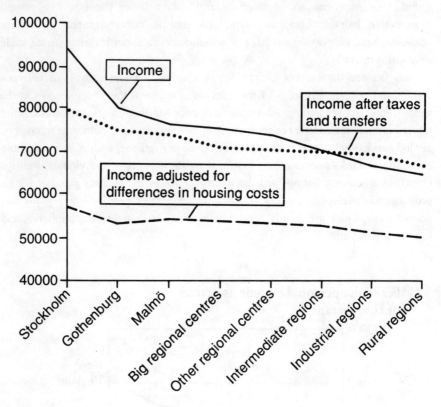

Fig. 3.15. Income and purchase power in Swedish regions 1990-91. Source: Statistics Sweden.

In the state of North Carolina (included in the Southeast in Figure 3.14), per capita personal income in 1986, when measured at the county level, ranged from USD 7.097 to USD 16.786, while in 1970, for the same 100 counties, the range was from a low of USD 1.668 to a high of USD 4.144. Meanwhile, and of critical importance to our concerns with regional issues, the correlation between per capita income and urbanization remained very high. So high incomes continue to be associated strongly with a high degree of urbanization (Gade, 1991).

The stability of the regional income differences in Sweden is striking in contrast to North Carolina (Figure 3.15). In terms of disposable income the average levels differ very little between various types of regions.

3.4 Focus on Intermediate Regions in the Swedish Geography

In Sweden, the municipality reform of the early 1970s resulted in 284 more or less functional units in terms of labour and housing markets. Increasing mobility in terms of daily commuting has eventually extended the local labour market areas beyond the administrative barriers. Hence, planning for the municipality is becoming less relevant than planning for the local labour market area, even if the legislative foundation for this is presently weak. Still, the underlying vision behind regional development in Sweden is the creation of coherent and functioning markets for labour, service distribution and housing. As a result, Sweden today can be divided into 111 functional local labour markets (Carlsson, et al, 1993). Each of these labour markets is characterized by intensive daily commuting. A cluster analysis of these labour market areas based on size and location, industrial structure and the qualification of the labour force, allows us to divide Sweden into five[2] dominant types of local labour market areas:

Metropolitan 1 (Stockholm)
Metropolitan 2 (Gothenburg and Malmö)
Regional centres (medium sized)
Industrial and Intermediate
Rural

Table 3.3 summarizes some current measures in each of these clusters, showing that the most urbanized and service-oriented regions also evidence the most rapid growth during the last decade. The income distribution is rather even, with only the Stockholm region showing a marked positive deviation from the national average. The equal distribution of public services is clear from the similar proportion of public service employment at different labour markets. In terms of migration (shown by the in-migration/out-migration ratio during 1986-89, referred to as 'preference' in Table 3.3) the most urbanized and service oriented regions are gaining. This is more evident for highly educated and young migrants, while families with children seem to have a preference for types of regions other than metropolitan. During the present extremely deep recession the distribution of unemployment is rather even, allowing only for a negative deviance in rural areas and a slightly positive deviance in Stockholm.

---

2 These five types of local labour markets are aggregates of ten clusters described in Chapter 4.

Table 3.3. Regional indicators: growth rate, unemployment, income, public service employment and in/out migration ratio, 'preference', for selected groups. Five types of labour market areas in Sweden. Source: Statistics Sweden.

Regional Indicators	Clusters of local labour market areas					
	Metrop 1	Metrop 2 + 3	Regional centres	Industrial and Intermediate	Rural	Total for Sweden
Employment, growth % per year 1980-90	1.7	1.3	1.3	0.8	1.0	1.1
Unemployment rate 1992 Oct index	90	98	99	99	133	100
Income/capita 1990 index	116	99	99	91	91	100
Per cent employed in public services	32	33	36	32	35	34
Preference young people in/out %	130	127	97	82	79	-
Preference families in/out %	38	73	133	127	140	-
Preference persons with post-secondary education	143	114	95	82	82	-
Total preference	107	114	102	85	94	-

We expect the regions outside metropolitan areas to function in different ways under postindustrial conditions. Their function as traditional industrial areas is overlapped and intermingled with their function as residential and recreational areas. "New" actors entering these spaces, including service producers within the information sector, retirees from metropolitan areas, settlers with specified housing preferences of a combination of rural and urban environment. They are replacing, supplementing and sometimes conflicting with the traditional industrial and agricultural activities and related lifestyle patterns. The new mix of actors is reflected also in a complex mobility pattern, both in the labour market and geographically.

With reference to the transformation process in rural areas, we define the concept of *"urbanized rural areas"*. The geographical pattern of the urbanized rural areas in Sweden is shown in Figure 3.16. This is a type of area where there is an acceptable population base for most types of daily basic services and an ordered - even if limited - wage labour market, although at the same time conditions are poor for advanced services for both households and businesses. This proposal concerns - under Swedish conditions - smaller central places or other population centres (2 000 - 10 000 inhabitants) with a commuting area (30 km) which is located relatively far away (>30 km) from bigger settlements. According to this definition, the urbanized rural areas are local labour markets that are found in most parts of Sweden and altogether contain 12 per cent of the total population. This consequently excludes remote areas with scattered settlements and with households which are almost completely dependent on their own enterprise or guarding of natural resources. The rural periphery covers a large land area, but very few people (approximately 2 per cent of the total population).

Data reveal that the urbanized rural areas are well equipped when it comes to basic services, both in terms of personal and of business services. On average these areas provide almost 80 per cent of what is generally found in established urbanized regions. However, when it comes to more specialised services, primarily for business purposes, the situation is different. The level of these services represents only one third of what is found in typical urban regions (Johannisson, et al, 1989). In the long run, this creates a problem for economic growth and renewal in these regions. The importance of the supply of producer services is more and more recognised (Cf. Sjöholt, 1990).

Labour markets in intermediate regions in Sweden are, in spite of their small size, able to offer employment at almost the same level as bigger regions. The average employment frequency is only a few per cent points lower than in big urban regions (Johannisson, et al, 1989). The differences in unemployment rates are not significant, neither in periods of high or of low level of unemployment.

The idea guiding regional and rural development in Sweden is the creation of coherent and functioning markets for labour, service distribution and housing. Public intervention and public service production are prominent elements of this idea. To some extent, however, this idea is now challenged by the emergence of what might be labelled the "arena society" (Johansson and Persson, 1991). This implies a slackening, at the household level, of the geographical links between the workplace, residence and place of education. This is reinforced by new models for labour organisation and improved infrastructure, i e for telecommunications. Internationalisation processes widen the scope for many firms - through expansion of

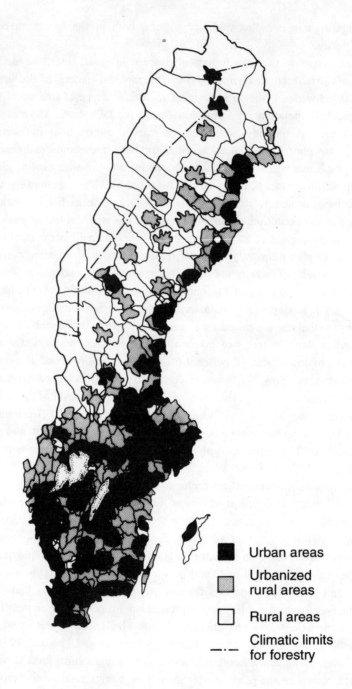

Fig. 3.16. Division of Sweden into three categories of regions. Source: Johannisson et al, 1989.

markets and partnership - as well as for many individuals - through international travel and education. Increasing standards of living and increasing stress on individual and private (rather than public) alternatives adds to mobility and flexibility. This has a general significance in all regions, but is especially noticeable and important in intermediate regions. More and more households are becoming less dependent on only local resources and local incomes. Some residents migrate to these regions in certain phases of the life cycle and the rationale for rural and small/medium town living has more to do with the perceived quality of life than with the quantity and quality of local labour demand and the local supply of services. Some of the new residents are more or less independent on local sources of income, partly as a consequence of the profession and the way to organise the work, partly because of the new communications technology.

The traditional function of the local labour market is, consequently, less important for many people, while dependence on the economy and the technology of other regions increases. This challenges the somewhat settled and static character of non-metropolitan areas that has, at least implicitly, been one image guiding the support system in all Nordic countries.

At the same time, planning and policy based on collectivism and solidarity is challenged more and more. Among the youngest generation there seems to be a movement towards increased individualism. There is a diffusion of postmaterialistic ideas and values, for example in form of environmental concern, partly from urban areas.

In addition, competition for labour and for locations in the metropolitan areas might lead to an increased interest among foreign companies in establishing plants beyond the edge of the central regions. Several towns in the intermediate region with reliable infrastructure in connection with a loyal labour force and with reasonable wage demands might be of interest for these companies. Many of these can offer an attractive environment for living and production, which is qualitatively different from big cities.

Finally, contraction of public services will not restructure the local labour markets and the service provisions in intermediate regions as much as in peripheral and central regions. Business services are expected to expand in intermediate regions, at least where infrastructure and housing conditions are favourable.

4 SPATIAL DIMENSIONS OF THE EMERGING KNOWLEDGE SOCIETY IN SWEDEN[1]

4.1 The Race for Increased Productivity

Over the last few decades an educated labour force (with qualifications at the post-secondary level) has become of considerable and increasing importance for the development of economic life and more diversified social structures in all advanced economies of the world. The share of more highly educated people in a local labour market not only indicates the qualitative level of existing economic activities, it is also indicative of general preconditions for the establishment or relocation of institutions and enterprises demanding well educated people.

However, it should be noted that the connection between education and productivity in terms of production per employee is not immediate and short term but rather long term. It is a complicated task to find efficient strategies to achieve a higher level of productivity. It is not just a matter of people's knowledge and competence. Productivity is a rather diffuse measure for success and future options. A capital intensive branch, including steel works or paper mills, often has a high level of productivity due to the intensive use of efficient machinery. Robots do a substantial part of the practical work. A stagnating branch may experience a rapid increase in productivity as a consequence of the closure of the plants with the most outdated production methods. Yet, the national branch as a whole may face problems competing successfully in an international context as a consequence of lower comparative advantages in total.

Even if there are problems identifying the role of education in detail there is a consensus among most researchers in various countries that there exist clear connections between education and economic development in a long term perspective. From the literature we have picked up some examples. Great Britain had a leading industrial position in global terms up to the end of the 19th century when Germany and the United States overtook it and achieved a dominant role in

---

[1] This chapter is largely derived from Axelsson et al (1994).

the world economy. One principal explanation for the loss of competitiveness among enterprises in Great Britain is the fact that the education system did not have a strong enough profile within fields of strategic importance for the renewal of industry. The opposite occurred in Germany. There explicit efforts were made - often in concert with leading industries - to strengthen opportunities for education within the fields of the technical and natural sciences. A similarly successful process commenced in Japan after the mid 1950s as a consequence of a conscious strategy to satisfy the demand for technicians and economists. The result was a boom that influenced many economies around the world. Japanese industry became very competitive globally, despite long distances to markets in Europe and the United States (Thurow, 1992).

In a short term perspective productivity is also linked to variations in trade conditions. During recessions productivity decreases, while it increases when the economy is strong. The positive development of productivity in the latter case is due to the way in which capacity, plants, machinery and stored goods, are taken back into operation. This causes a rapid impact on production without creating a critical need for more employees. In the opposite case - during recessions - employment rarely decreases as much as production volumes. Employers cannot and will not dismiss personnel as a direct reaction to decreasing demand for their products, as they want to maintain the levels of competence among their staff.

The diffusion of formal knowledge among different types of local labour markets is a complicated process involving a variety of pull and push factors. The provision of highly educated people is dependent on the geographical dispersion of higher education opportunities and the variety of educational facilities available in different locations. As people with an extensive formal education represent a mobile resource, it is also a question of the character of the reception structure and its competitive strength when it comes to attracting economic activities which demand personnel with extended formal training. In- and out-migration flows will either drain off qualified people or act to upgrade the local level of qualifications.

The distribution of government resources for higher education in Sweden favours quite concentrated geographical effects. There are 7 cities with universities which offer postgraduate programs leading to a doctoral degree. Universities and specialised higher education institutions in these cities were responsible for 77 per cent of the total number of full-time equivalent annual study places in the financial year 1991/92. As part of a university reform in 1977 small and medium-sized university colleges were established in cities all over Sweden to promote greater geographical dispersion. There are now 22 cities which have university colleges. In another 20 cities and towns decentralised education is offered. There are also distance education programs in operation.

However, the share of government resources distributed outside the universities is small. For example, the university colleges receive less than 10 per cent of that which is transferred to the universities. A great number of sparsely populated municipalities, which together have almost the same total population as the three metropolitan regions, receive almost no resources for higher education.

In the Swedish qualification landscape significant local supply effects appear at the post-secondary level. The local labour markets belonging to the upper quartile in terms of the proportion of people with more than two years of university studies is to a large extent reflective of the location pattern of universities and university colleges (Figure 4.1). However, the general importance of geographical proximity to cities with higher education institutions seems to be low. Significant magnetic effects appear in some municipalities lacking higher education facilities but with a specialisation in industries demanding a large share of more highly educated people, especially technicians.

Table 4.1 Share of employed people who have qualifications at the post-secondary level (> 2 years) in different sectors in 1985. Source: Statistics Sweden.

Sector	Marginal regions in the north[2] %	University city of Umeå %	Average for Sweden %
PRIMARY PRODUCTION	1.3	4.9	1.8
MANUFACTURING	1.1	3.1	4.0
- raw material based	1.1	4.2	3.7
- consumer goods	1.2	1.1	1.9
- investment goods	1.2	2.5	4.7
CONSTRUCTION	0.6	2.7	2.0
TRANSPORTATION	0.5	3.0	2.9
BUSINESS SERVICES	4.8	9.4	10.2
HOUSEHOLD SERVICES	3.2	6.0	4.9
EDUCATION	45.5	50.6	51.3
HEALTH CARE	4.2	10.5	7.8
PUBLIC ADMINISTRATION	6.3	14.4	13.7
TOTAL	5.7	13.5	9.1

---

[2] Sparsely populated municipalities in the six northernmost counties.

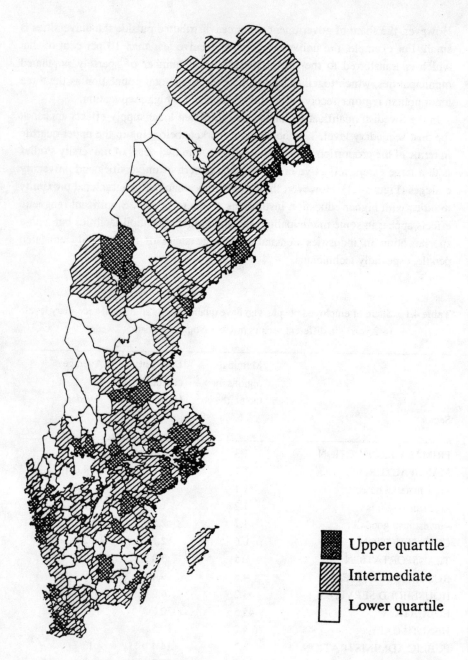

Fig. 4.1. Density of people with post-secondary education (>2 years) in Sweden 1985. From (Johansson et al, 1991).

Among different sectors of the economy we find great variations in the demand for people with qualifications at the post-secondary level. Table 4.1 illustrates the share of more highly educated people in various sectors and the regional disparities between a university city and marginal regions. The definition of marginal regions used here is municipalities with a small population and with long and inconvenient conditions for daily commuting to cities with universities. We find the most pronounced disparities between marginal regions and the rest of the country within the private sectors of the economy. However, also in the young university city of Umeå there appear in some sectors lower shares of more highly educated people than average for Sweden. From this we may conclude that a local university is no guarantee for a high local supply effect. The reception structure of various sectors creates limits.

It has been and still is an ambition of Swedish regional policy to provide people with a good social service level wherever they prefer to live in the country. Qualified services such as modern medical attention and health care, education and administration demand people with certain minimum levels of education. These spatial and sectoral demands are to some extent regulated by the central government, which has commanded the municipalities to offer certain types of services above certain minimum levels of quality. Within the public sector some types of jobs are also closely linked to certain levels of education and its content. Examples are medical doctors, dentists, psychologists and teachers. Almost no correspondingly strict rules exist within manufacturing and the private service industries. As a consequence of this today we find almost two thirds of the people with a higher education in Sweden within the public sector - a sector which only represents one third of the whole labour market. There is also a distinct pattern of concentration of people with the longest higher education to public sector services and administration. In all types of regions - metropolitan, intermediate and marginal - a similar pattern appears. The most extreme examples are found in small sparsely populated municipalities where we can find that public sector activities employ up to 80 per cent of the people with a higher education. The dominance of the public sector as an employer for academicians means that the general educational level in a municipality is a consequence of the relative size of the public sector. Figure 4.2 illustrates the formal qualification profile of various employment sectors in Sweden.

In terms of the spatial distribution of the more highly educated people working in the private sector, there is a very high concentration. For example, among people in Sweden with more than three years of higher education working in the private sector, we find 42 per cent in the metropolitan region of Stockholm.

Another aspect of the regional disparities is the formal preparedness for taking university courses. In marginal regions the share of those employed with

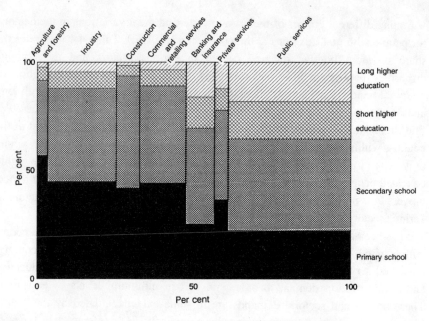

Fig. 4.2. Labour force by sector and level of education. Source: Statistics Sweden, FoB 1990.

qualifications at the secondary school level (> 2 years) was approximately 6 per cent in 1985, while the corresponding share in university cities was approximately 11 per cent.

A detailed analysis of the stability and balance between the inflow and outflow of people with a university degree during the latter part of the 1980s has been carried out for local labour markets in the two northernmost counties of Sweden - Norrbotten and Västerbotten - and with Stockholm as a metropolitan reference. The local labour markets in the two counties have been classified into three groups: regional core, intermediate and peripheral areas. We find a decreasing pattern of stability among people with a university degree from the metropolitan region to the periphery (Figure 4.3). By stability we mean the number of permanently employed people within the same local labour market during the period 1986 - 1989, divided by the number of people employed in the same category in 1986. All examined types of local labour markets had a positive net expansion of more highly educated people with the highest values for the regional cores and Stockholm. However, in the age group 35 - 50 years the net changes were close to zero for all types of regions examined.

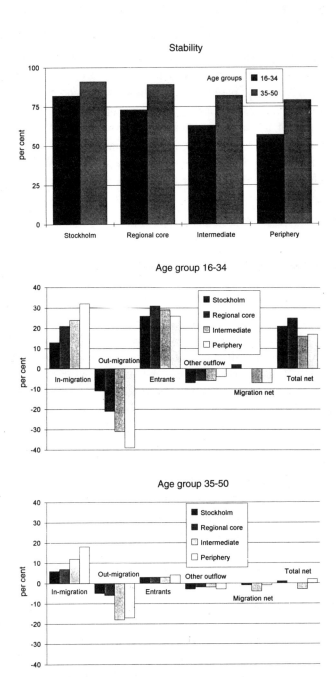

Fig. 4.3. Geographical stability and flows of people with certificates from higher education institutions during 1986 - 89. Stockholm compared with the two northernmost counties in Sweden. Source: Statistics Sweden.

Even if there was a positive development in the small peripheral local labour markets there was an even stronger expansion in the larger local labour markets. Within the framework of general expansion we find widening regional gaps. The present situation is that compared to metropolitan regions and most medium-sized cities, the marginal regions are lagging behind by approximately 10 years in terms of the proportion of labour with a short post-secondary education and by more than 20 years in terms of a longer post-secondary education.

One obvious conclusion is that there has to be a very strong reallocation of resources to create a distribution system and/or a reception structure for people with a higher education which may more than marginally slow down the concentration trend.

4.2 Characteristics of the Knowledge Society

The knowledge society gained a position of prominence in politics and planning during the 1980s. However, this does not mean that it is a phenomenon that appeared then for the first time. In fact it is based on a cumulative process over several decades that consequently becomes more and more visible as its implications and effects are gradually spread to all parts of social life. In most industrialized countries the share of service professions became greater than that of manufacturing professions during the 1950s and 1960s. However, the growing dominance of service based professions did not immediately result in a change from an industrial society to a knowledge society. In a labour market dominated by manufacturing production most of the labour is occupied with the construction, assembly, maintenance and transportation of goods. Even in a service dominated labour market a great share of those employed have corresponding tasks. Often most of these types of jobs do not require people with any special educational qualifications.

Even future oriented estimations of labour market demand show that a great share of the labour force is expected to handle knowledge and information in the context of routine production. According to a division of occupations within information handling industries into three groups only one group, representing approximately 20 per cent of the employment, demands people with a higher education (Reich, 1992). The two groups with a low demand for qualified people are the routinized handling of information and person-to-person services. In the first category we find jobs which are labelled "back office jobs". This is a growing category of occupations responsible for handling continuously growing flows of information and transactions.

The explicit knowledge oriented occupations fall into a category that Reich (1992) has labelled symbolic analyst. Even if the group plays a critical role for the creation of new resources and opportunities in the economy it represents a small share of the labour market and there is a complicated barrier to any significantly more rapid increase. In Figure 4.4 some examples of the way in which the knowledge society creates new types of occupations which are based on combinations of specific fields are given. In the Figure we may find combinations such as process-designer or planning consultant, as well as even more complicated combinations such as project-development-coordinator or business-strategy-adviser.

One question, which is often raised, is whether or not the knowledge intensive part of the labour market demands specialists. Figure 4.4 supports the argument that there is a growing demand for people who can combine knowledge from more than one specific field. To maintain and produce further competitive strength an advanced economy must be active in finding new approaches for handling the inefficiency and problems which appear.

The spatial distribution of knowledge based occupations will be more and more important as a locational factor. The winners in this game will only be those places or regions in a nation state which have the ability to provide an efficient educational back-up and to create a creative milieu where the competence may be used efficiently. An on-going renewal process within the education system is of critical importance. This demands intensive and well organized contacts with leading research centres around the world. Those regions and places which face problems in competing successfully in this international context will not have such good opportunities to develop a knowledge based industry of significance. Instead their role will be focused on routinised service production or person-to-

Communications	Management	Engineer
Systems	Planning	Director
Financial	Process	Designer
Creative	Development	Coordinator
Project	Strategy	Consultant
Business	Policy	Manager
Resource	Applications	Adviser
Product	Research	Planner

Fig. 4.4. Combinations of fields within knowledge based occupations. Source: Reich (1992, p. 183).

person services. In this respect the spatial division of labour within the knowledge society conforms with the corresponding character of the industrial society, where routinised manufacturing plants are located to less developed countries with cheap labour and knowledge intensive plants to the most developed countries.

We may conclude that the most important geographical characteristic of the knowledge society, compared with the industrial society, is that the role of national economies is decreasing due to the advantages that single places or regions may achieve as a consequence of active participation in international networks. The outcome of this is that new types of regional differences are appearing. This new spatial pattern is related to critical mass, local concentration and global networks rather than regional diffusion. Today this new pattern is often very clearly demonstrated among export oriented industries. They have a dualistic geographical orientation with a strong home base strategy combined with global market penetration. In the local context the accessibility to qualified labour often forms a very important resource.

The driving forces behind changes within the industrial society had a strong bottom-up character; organized labour pushed for increasing influence on work conditions. The expansion of education focused on the spread of basic knowledge to all inhabitants, which corresponded well with competence demands from the expanding industries. Historical analyses have stressed the importance of the labourers' ability to read as a major factor behind the strong economic growth during the past 100 years. After the initial stages of industrialism an important prerequisite for further economic growth was a spatial and social diffusion of welfare. In this way consumption could expand and stimulate the home market oriented enterprises. Thus strong interdependencies were created between mass production and mass consumption. A key role in these processes was also played by some very successful commercial innovations. In most cases they seem to be derived from work done by individual innovative people rather than the products of organized collaboration between people with certain educational experiences.

In many respects the knowledge society has a rather different character. It is demands for further knowledge among people in knowledge based occupations which form the critical conditions for competitiveness and economic growth. While mass production for national markets favoured the equalisation of wage levels, the knowledge society fosters widening gaps between the labour market for knowledge based occupations and that for routinised service jobs. The migration pattern among people with knowledge based occupations shows a positive immigration ratio for developed countries with significant differences in wage levels between various types of knowledge based occupations (Reich, 1992). In the recent Swedish debate and policy making process there have also been recommendations to increase wage levels for key professions in order to stimulate more people to undertake some form of higher education (Henrekson, 1993). It has also been suggested that the wage levels should be decreased for people with little formal education. Similar attitudes are found within private industry. The widening of competence based gaps in wage levels is regarded as an instrument to stimulate initiatives, creativity and interest in education.

We may conclude that two basic problems exist for policy-making and planning as a consequence of the character of the knowledge society. The first is the critical role of a steady increase of productivity within the labour markets of the knowledge intensive occupations. The other is the great social challenge to find new ways for transferring welfare to people with a lower level of education (Drucker, 1993). Both of these problems have a significant spatial dimension.

Besides the discussed division of the labour market cyclical and structural processes appear causing problems in the form of unemployment, under-employment and over-employment. One of the structural reasons for unemployment is the removal of production to other countries as a consequence of strengthened price competition. Another structural factor involves technical investments which reduce the need for man-power. There is also a tendency towards increased under-employment among people who want to work full time, but cannot find that type of job. Simultaneously there is a growing number of people who work more than 60 hours a week. Over-employment is growing particularly among people in knowledge based occupations where the production result is mainly a consequence of factors other than the time spent at the work site (Schor, 1992).

In all parts of the developed world it is expected that a considerable increase in the number of highly educated young people will occur during the coming years. Estimations in various countries show a similar pattern of expectations in this respect, even if there are problems comparing countries due to different definitions. For example, the calculations for Germany seem to be very optimistic. The share of highly qualified people there is expected to increase from 25 per cent in 1987 to 37 per cent in 2010 (OECD, 1993).

If the occupational structure remains in its present shape this will mean that a growing number of people will do the same type of job as earlier but with a higher educational background than their predecessors. However, we may also anticipate changes on the demand side, characterized by an increased demand for people with a high level of formal qualifications.

In many of these countries there is a discussion of the needs and possibilities to increase the demand for a more highly educated labour force within industry. In most cases the average share of labour with a higher education is approximately 5 per cent. However in a few branches this share is much higher. A prerequisite for a considerable increase of highly educated people within industry is a strong growth of those kinds of industries. Another important argument for a larger share of people with a higher education within the private sector is that the demand will not be as strong as earlier from the public sector organisations. In many countries besides Sweden there is now a clearly dominant public sector orientation among more highly educated employees. For example, the share is 69 per cent in Austria, 60 per cent in Belgium, 66 per cent in Germany but only 48 per cent in the more market oriented Canada (OECD, 1993). A considerable change in the demand structure from the public to the private sector represents no conflicts of interests and priorities in these national contexts. In fact, there is already a tendency in Sweden, as well as in other advanced welfare economies, towards decreasing demand from the public sector.

The expected character of the labour markets of most western countries when we enter the 21st century is full of contradictions. Most actors agree that economic growth demands a higher level of competence among the labour force. On the other hand the demand for highly educated people is still very low within most industrial branches. This leads to the following two conclusions which are of relevance for education policy and regional policy. First, there is a push for a structural transformation including closures or the relocation of less productive enterprises and plants to countries with lower costs. At the same time industries with a qualified form of production are expanding and in this case opportunities are being taken to locate the expansion to regions with a good supply of competent labour. Second, parts of the education system must be transformed so it will more closely fit the demand among industries with a growth potential.

4.3 Fragmentation Tendencies Among Swedish Local Labour Markets

Does the knowledge society, with its increasing focus on the production, movement and adjustments of information rather than the production of goods, mean that the regional context in the form of an integrated and well defined geographical territory for working and living is becoming more difficult to define? We can see a tendency towards a more flexible character of the daily arena and related commuting patterns for a growing number of occupations. However, the transformation towards the dominance of this new behaviour is a rather slow process, which means that for a rather long time into the future it will still be relevant to identify functional local labour markets based on traditional, rather stable commuting patterns between homes and working places.

In the present situation with high unemployment, increasing differences in wage levels between people with a short term and long term education and a stagnating - and in some cases a decreasing - public sector increasingly visible qualitative differences are appearing between various local labour markets. A classification based on population figures for various local labour markets shows an unbalanced regional pattern. If we take the significant flows of commuters across borders of municipalities into consideration, the local labour markets in Sweden vary in population from 2 000 to 950 000 people (Carlsson et al, 1993). There is a significant bias in the distribution. Almost 80 per cent of these 111 defined local labour markets (LMAs) constitute just over 1 million people in the work force, which is approximately one fourth of the total national labour market (Table 4.2). This group of local labour markets, each with less than 50 000 people in employment and many of them much smaller, are here labelled "small" and "medium-sized". They have a rather different qualification profile among their labour force than the bigger local labour markets. As is shown in the Table, a majority of the more highly educated people are working and living in the 25 biggest local labour markets. In the qualification structures we can identify a growing tendency towards a polarisation between the limited number of big local labour markets and the great number of small local labour markets. Figure 4.5 further illustrates that the curve representing the distribution of more highly educated people is steeper than the curve representing the distribution of employment among the local labour markets.

Table 4.2   Local labour markets (LMAs) in Sweden after size, share of employment and level of higher education in 1990. Source: Statistics Sweden.

Size of LMA	Number of LMAs	Average employment	Share of total employment in Sweden, %	Share of all with higher education, %
Small (< 10 000)	43	5 000	5	3
Medium-sized (10 000 - 50 000)	43	20 000	19	13
Big (> 50 000)	25	130 000	76	84
Total for Sweden	111	40 000	100	100

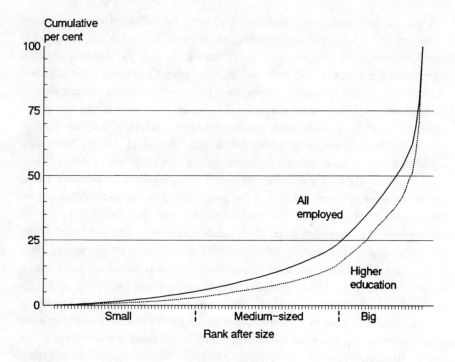

Fig. 4.5. Local labour markets in Sweden ranked according to number of people employed in 1990. Cumulative per cent of total employment and those with a higher education. Source: Statistics Sweden.

Factors other than population size also explain the qualitative differences between various labour markets. A more careful description of the regional structure of the emerging knowledge society may take into consideration the formal qualification structure among the local labour force, the distance to the nearest centre with education facilities, the structure of private and public employment and organisational frameworks. Among the local labour markets we may identify clusters of local labour markets characterized by a high degree of similarity in these respects. These clusters give us an overview of the geographical differences in the functional criteria of local labour markets. It provides a way of mapping vulnerability, development potentials and capacity, in order to offer flexible options for various types of labour.

In a study by Statistics Sweden and the government's Expert Group on Regional Studies a cluster analysis was undertaken based on the types of factors mentioned above (Carlsson, et al, 1993). The following four types of factors were chosen to indicate various critical characteristics of local labour markets in the perspective of a growing knowledge society: (1) The total share of more highly educated people and the specific share of people with a higher education within the fields of the technical and natural sciences, (2) A measure of the density of local supply and demand (population and territorial area of the local labour market), (3) The economic structure of the local labour market, and (4) The organisational structure of the enterprises. All of these factors were given the same weight in the cluster analysis. This resulted in ten clusters of local labour markets. The local labour markets are located in various parts of the country except for the metropolitan areas, which form three separate categories (Figure 4.6). Table 4.3 lists the labels derived if we use the most characteristic feature of each cluster. Category 6, labelled "small and medium-sized with a diversified economic structure", may be regarded as the most typical of the intermediate regions. Approximately one third of all local labour markets belong to that category according to this cluster analysis.

Table 4.3   Description of ten categories of local labour markets (LMAs) in Sweden.

Category	Number of LMAs	Share with more than 2 years of higher education
1. Metropolitan area of Stockholm	1	14
2. Metropolitan area of Gothenburg	1	12
3. Metropolitan area of Malmö	1	12
4. Regional centres	25	9
5. Centres in the north-western peripheral areas	5	8
6. Small and medium-sized with a diversified economic structure	34	7
7. Dominated by large scale industry	10	6
8. Dominated by small scale industry	11	5
9. Rural LMA	18	6
10. Small LMA dominated by a single industry	5	5

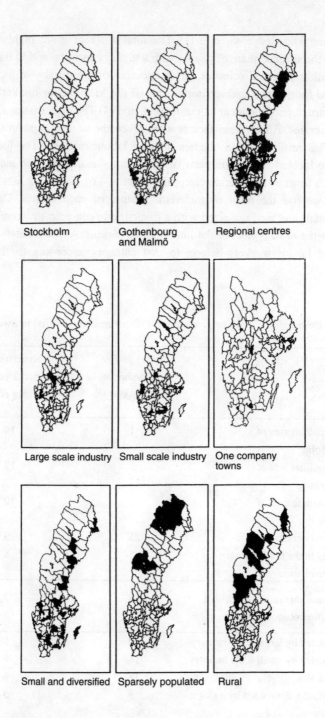

Fig. 4.6. Categories of local labour markets in Sweden. Source: Carlsson et al 1993.

We may draw the following general conclusion. At the present stage in the process towards a knowledge society there is an obvious fragmented regional structure. Among the ten different categories of local labour markets identified, the most distinct demarcation line is found between a limited number of large urban centres with a high level of qualifications among the employees and a geographically dispersed category of small labour markets with much lower qualification levels. The results underline our general hypothesis about growing differences - or fragmentation - among Swedish local labour markets.

4.4 Patterns of Restructuring Labour Market Areas

In the 1980s, the national labour market expanded strongly in Sweden. The total number of jobs increased by 400 000, which is almost 10 per cent. But during the first few years of the 1990s, unemployment has increased dramatically and the structure of the labour market has been clearly changed. Several sectors of the economy are now in a phase of a thorough restructuring process which involves public services such as health care and social services, private services such as financing and banking as well as several branches of the manufacturing industry. The structural adjustment processes influencing today's labour market are partly the results of events during the last decade. These processes will have strong regional consequences. In a regional perspective, the growth period - primarily covering the period 1982-90 - varied strikingly between different types of local labour markets. Figure 4.7 illustrates the different growth rates in 10 categories of local labour markets (LMAs) in comparison to the country as a whole.

The metropolitan labour markets of Stockholm, Gothenburg and Malmö had growth rates well above the national average. This is also true for the regional centres. Altogether these four categories of LMAs account for approximately 3/4 of the total labour market in Sweden. Small labour markets, whether dominated by manufacturing industry or more service oriented activities, had growth rates less than one half of those recorded in Stockholm and Gothenburg. The same growth rate was recorded in the regions in the northwestern interior of Sweden; i.e, the area considered to be the prime target for regional policy. Small rural labour markets throughout the country were the least expansive. A pattern of duality in the growth rate - between the few large labour markets and the numerous small ones - seems to have emerged during the 1980s.

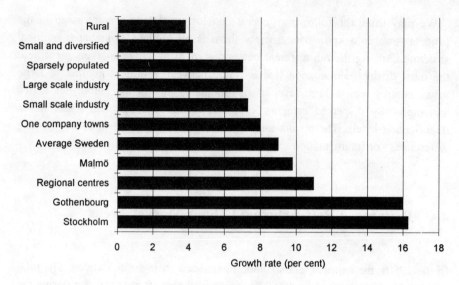

Fig. 4.7. Growth rate of total employment in categories of LMAs in Sweden 1980-90. Per cent. Source: Axelsson et al, 1994.

We observe that the structural situation in 1980 was important for the pattern of growth during the following 10 years. In Figure 4.8, the categories of LMAs are

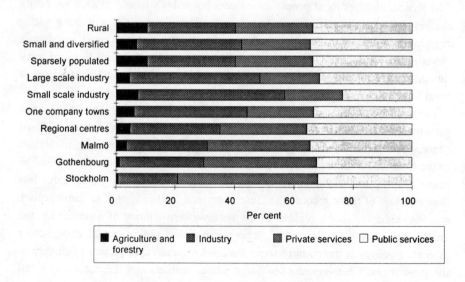

Fig. 4.8. Structure of the labour market - four sectors, 1980. Categories of LMAs ranked according to total growth rates 1980-90. Source: Axelsson et al, 1994.

ranked according to their total growth rate between 1980 and 1990. Labour markets with a comparatively large proportion of activities in either agriculture and forestry or the manufacturing industry ranked lower in the total growth rate of jobs. Deindustrialisation was not fully compensated for by growth in the service sectors.

Small local labour markets also had a reduced growth rate in generally expansive branches; for example, in private services such as banking, insurance, retailing and business services. In Stockholm, these private services accounted for 50 per cent of the total growth, while they accounted for only 10-20 per cent of the growth in small local labour markets. In addition to this, Figure 4.9 shows that deindustrialisation is a dominant phenomenon in most small labour markets.

Public services still experienced major expansion in the 1980s. In all labour market areas, except Stockholm, growth was considerable. For big labour market areas such as the two other metropolitan regions and the regional centres, public services accounted for 50 per cent of the total growth. In small labour markets, public services accounted for more than half of the new jobs. Without this public service expansion, most small labour markets would have experienced a net decrease in the number of jobs.

In the 1990s, industrial development has again come into the focus of the debate on regional and national economic development. The decline in the number of

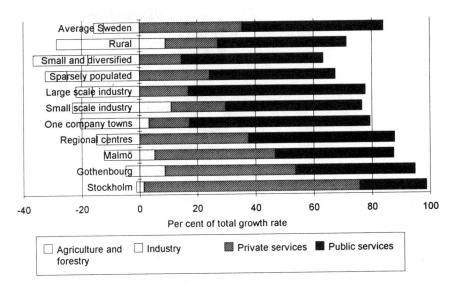

Fig. 4.9. Employment change by sector 1980-90. Per cent of the total growth rate in categories of LMAs. Source: Axelsson et al, 1994.

industrial jobs is used to illustrate the precarious situation of the country as a whole. It is sometimes argued that the industrial base is eroding.

In order to describe the performance of the industrial sector at local labour markets a functional classification of branches has been used.[3] The manufacturing industry is classified according to the dependence on production factors such as capital and labour as well as research and development. In addition, there is a class of branches labelled "sheltered", defined as such if foreign inputs are less than 10 per cent and exports are also less than 10 per cent. This means that important parts of the food, forest and building material industries form the core of the sheltered branches. Figure 4.10. reveals a relatively fragmented spatial pattern of development in these functional branches during the 1980s. Capital intensive and labour intensive industry were responsible for the largest relative decreases of jobs. This is because of the structural transformation of raw material oriented branches within the food and forest industries as well as metallurgy. Another set of branches showing decreasing labour demand was the knowledge or technology intensive industry, which was the largest class with more than 27 per cent of industrial employment in 1990. This set of branches largely consists of the

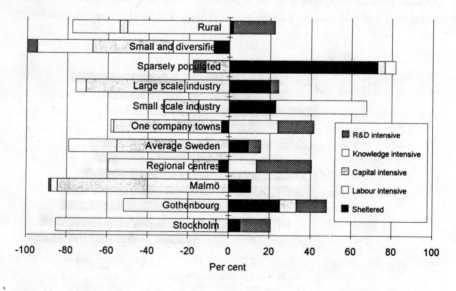

Fig. 4.10. Employment changes in categories of LMAs in Sweden 1980-90. Sets of branches, contributions to changes in total manufacturing industry. Source: Axelsson et al, 1994.

---

3 Compare Table 3.1.

manufacturing of fabricated metal products, machinery and equipment; they are oriented towards investment goods. It is worth noting that the largest absolute and relative decreases in the number of jobs in this set of branches took place on big labour markets where supplies of labour with a higher education are the biggest.

Sheltered industries have shown some growth on large labour markets such as Malmö and Gothenburg, but also on some small rural labour markets. The future of this home market oriented industry depends on the degree to which it can maintain its position on the Swedish market. Changing competitive conditions as a response to international free trade agreements mean that this set of branches will experience considerable restructuring.

A high growth rate is reported for R&D-intensive industry. This branch is the smallest with less than 10 per cent of the total employment within the manufacturing industry. Of the categories of LMAs, the growth rates in the R&D based companies are the largest in Malmö, Gothenburg and on small rural labour markets. However, in the latter, the growth was limited in terms of the absolute number of jobs. It is worth noting that Stockholm, which is assumed to have an important role as a location for R&D-intensive industry, had a low growth rate in this industry in the 1980s. However, 35 per cent of the employment in the manufacturing industry within this branch is already in Stockholm. Low growth in the last decade can be interpreted as a sign that Stockholm's R&D-industry is becoming mature in relative terms and is starting to deconcentrate.

To sum up, there are a number of observations about the spatial pattern of increasing labour demand during a period of economic growth which we wish to highlight. A large proportion of growth during the 1980s took place on the biggest labour markets such as the metropolitan areas and regional centres. For the one fourth of all jobs located at small labour markets there was growth as well, but it was far below the national average. If expansion had not taken place in health and social care, employment growth would have been weak or even negative in these small labour markets during the 1980s.

Small labour markets with a high proportion of employment in the manufacturing industry and agriculture at the beginning of the 1980s experienced a slow growth rate during the following decade. Within the framework of a total decline in industrial employment there was a strong increase in research intensive industry. The regional pattern is fragmented, however. The number of jobs in private services - mainly banking, insurance and consulting - increased mainly on bigger labour markets.

In the beginning of the 1990s a labour market which is clearly divided in relation to at least three factors - geography, level of education and gender - developed in Sweden. A sharp demarcation line divides a number of urban regions with a high level of formal education and a geographically dispersed group of small local labour markets with a much lower level of education. This concerns small labour markets with a uniform as well as a relatively differentiated branch and firm size structure. The gender pattern means that females with a lower level of education are overrepresented within public services and person-to-person services, while men are dominant in goods handling activities. This gender pattern becomes less marked as the level of education increases.

4.5 Migration Patterns

This section aims to answer three questions on the basis of empirical research about the effects of age, gender and level of education of the labour force in relation to migration.

- What is the impact of regional characteristics such as the industrial structure and the location of higher education facilities on the migration pattern of different categories of labour?
- In which ways do different types of labour market areas function as transmitters and receivers of migrants with a brief or extended formal education?
- What are the consequences on the size and direction of migration flows of different categories of labour choosing regions to work in and for residential purposes?

First of all it should be remembered that migrants constitute only a small group compared to those who are geographically stable. Over a period of four years, the number of migrants (moving across the border of a labour market area) corresponds to only 5 per cent of those of an economically active age who are stable. In addition it seems that the frequency of long distance migration has decreased in recent decades (Bengtsson & Johansson, 1992). While those with a university degree (more than two years) represent about 10 per cent of the labour force, they constitute almost 25 per cent of all those migrating between labour market areas. Migrants with a higher education - as in other groups - are strongly concentrated to those between the age of 20-30. However, an inventory of some 30 different labour categories (in terms of age, education, sector, marital status, children, gender) shows that young persons with a university degree are not the most likely to migrate. Young females between 16-24 years of age working in public services and without a higher education are more mobile.

Figure 4.11 illustrates that the tendency to migrate decreases with age. In higher age groups, the educational level does not have any impact upon the frequency of migration. Altogether, the three groups in the Figure - persons with a higher education, male industrial workers without a higher education and females active in public services without a higher education - constitute approximately 40 per cent of all the migration between local labour markets in Sweden. After 45-50 years of age most of them become stable within their labour market area. Compared to 100 people with a stable residence and a higher education between the ages of 50 and 65 years, there are less than four who move across a border of a labour market area within a time period of four years. Among those without a higher education in the same age group there are even fewer migrating. Most stable are the least educated males. On the average, 98.6 per cent of the latter remain in their region of residence for a period of four years. These results underline the totally dominant role of young people in migration in Sweden.

The high rate of mobility among young people is explained by numerous factors, such as education, work and the establishment of a family. There are also less rational motives. For example, the high rate of mobility among young females working in public services is not fully explained by job searching activities and

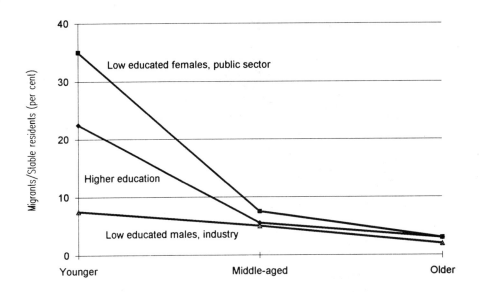

Fig. 4.11. The number of migrants related to the number of stable residents. Three categories of labour and three age groups for each. 1986-89. Source: Axelsson et al, 1994.

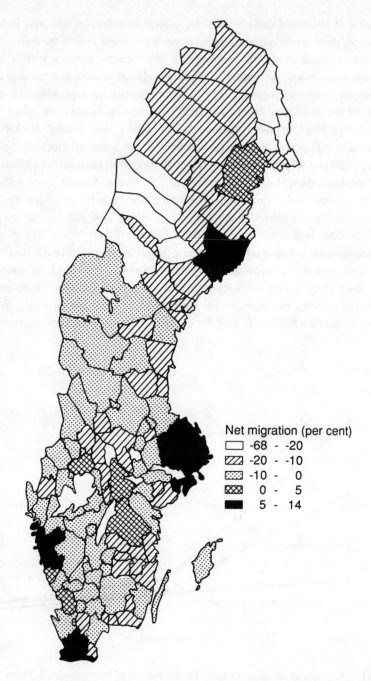

Fig. 4.12. Net migration between local labour markets 1986-89. Age group 16-24 years.
Source: Axelsson et al, 1994.

education. Many of these females seem to prefer big or medium sized localities both as a place for work and as residential location.

The necessity of migration to the few places with higher education facilities leads to an out-migration of people between 16-24 years of age from most regions without such facilities. This is not always looked upon as a problem, as long as there is enough return migration. Only university towns experience a positive net migration of people in the age group 16-24 years (Figure 4.12). Hence, the location and capacity of higher education facilities plays an important role in the mobility pattern of young people. Regions with large rate of out-migration - some labour markets in the north (except for the university towns), the old industrial region of Bergslagen in the middle of Sweden, and in the southeastern part - all have a limited capacity for higher education. In addition, employment opportunities in these regions are limited, which makes them a problematic choice for young people.

Now, let us analyse this detailed spatial pattern in terms of the earlier introduced 10 categories of LMAs. It turns out that net migration by young people between 16-24 years of age is very different in different categories of LMAs (Figure 4.13). All categories of LMAs except the three metropolitan regions have negative net migration in this group. The metropolitan regions benefited from extensive education capacities and excessive labour demand in the 1980s. Regional centres

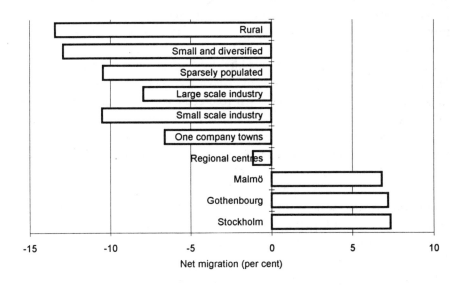

Fig. 4.13. Net migration of the age group 16-24 years. Per cent of the group's total in each category of LMA 1986-89. Source: Axelsson et al, 1994.

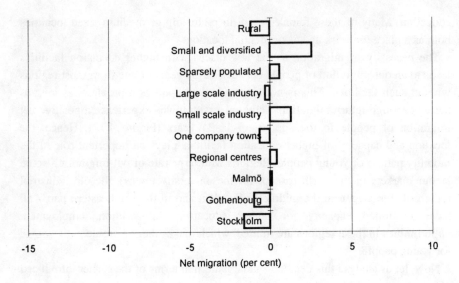

Fig. 4.14. Net migration of the age group 55-64 years. Per cent of the group's total in each category of LMA 1986-89. Source: Axelsson et al, 1994.

have a more terminal character, i.e., young people pass through them during their education. This is shown by the almost balanced out-/in-migration ratio. The pattern shown in the Figure seems to be geographically bimodal, i.e., young people reinforce a geographical division of Sweden. Again, the regional centres function as points of balance.

The decreasing geographical mobility by age is often explained by the increasingly established position at the labour, housing and marital markets. A small increase in migration frequency is recorded around the regular pension age (65 years), but it is well below the frequency in some other countries. Retired people often have contrary migration flows to young people. This means migration to labour market areas in "Small Town Sweden". Migration at this age is associated with return migration or the search for an attractive place of residence. Migration between categories of LMAs (Figure 4.14) shows that small towns without large scale industry are favoured. A separate analysis shows that elderly people with a higher education tend to stay in metropolitan regions to a larger extent. The spatial pattern on the level of labour market areas is shown in Figure 4.15.

People without a higher education form a heterogeneous group. We have chosen to show the behaviour of two subgroups: males in the manufacturing industry (Figure 4.16) and females in public services (Figure 4.17). Employment increased

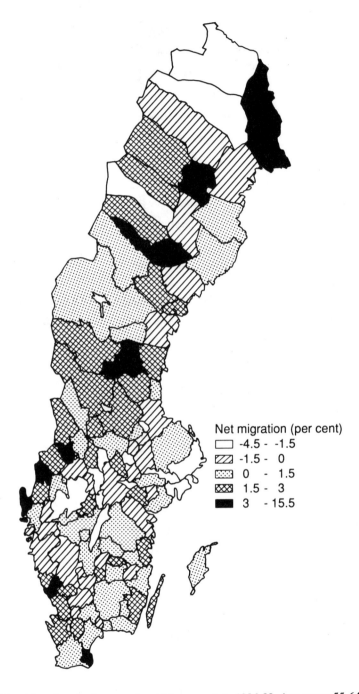

Fig. 4.15. Net migration between local labour markets 1986-89. Age group 55-64 years. Source: Axelsson et al, 1994.

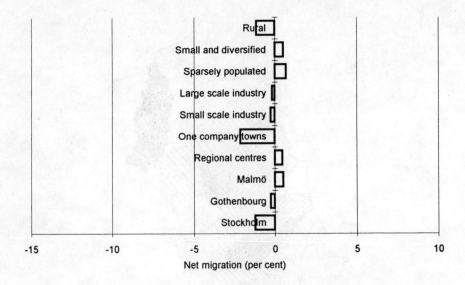

Fig. 4.16. Net migration, males without a higher education in manufacturing industry, 16-64 years. Per cent of the group's total in each category of LMA 1986-89. Source: Axelsson et al, 1994.

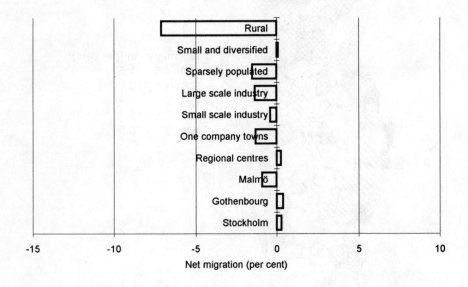

Fig. 4.17. Net migration, females without a higher education in public services, 16-64 years. Per cent of the group's total in each category of LMA 1986-89. Source: Axelsson et al, 1994.

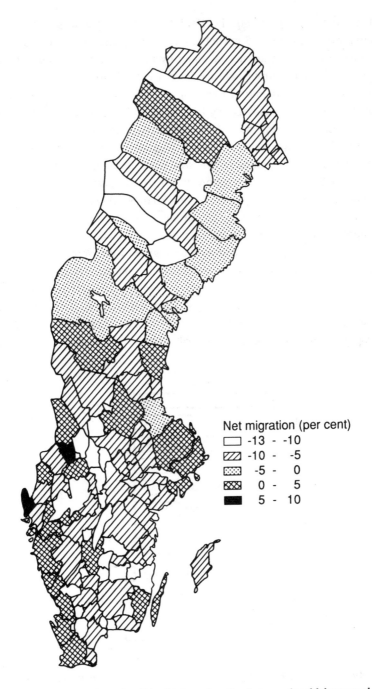

Fig. 4.18. Net migration of people with a higher education between local labour markets 1986-89. Age group 16-64 years. Source: Axelsson et al, 1994.

in both these sectors during the 1980s. Females in public services only had a positive net migration to Stockholm, Gothenburg and the regional centres. Rural and industrial regions are not favoured by this group. On the other hand, males working in manufacturing industry have a net out-migration from metropolitan regions except for Malmö. There is a net in-migration to regional centres and small labour markets without any large scale industry. Compared to the females, males in manufacturing do not "escape" from rural regions at the same rate. Persons with a higher education largely choose to migrate from the north and the southern interior to big cities and the west coast area (Figure 4.18).

Those with a higher education prefer to migrate to the labour markets employing the largest proportion of this kind of labour, i.e., the biggest labour markets (Figure 4.19). During the strong economic upswing in the 1980s the metropolitan regions were demanding labour at all qualification levels. Among the types of regions supplying metropolitan regions with labour we can distinguish rural labour markets and also the very sparsely populated regions in northwestern interior. It is evident that this drainage in terms of net out-migration contributes significantly to the fact that the level of education increases very slowly there.

The most important export regions for people with a higher education to metropolitan regions are the regional centres and regions with small scale

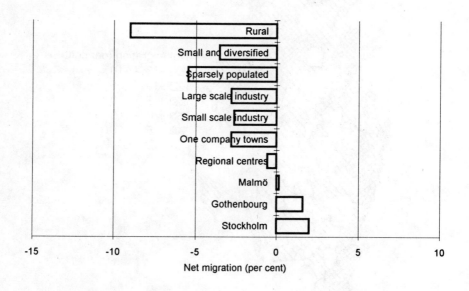

Fig. 4.19. Net migration, people with a higher education in age group 16-64 years. Per cent of the group's total in each category of LMA 1986-89. Source: Axelsson et al, 1994.

industry. These two categories of LMAs account for almost two thirds of the migration to Stockholm. Rural labour markets - with relatively high losses - contribute very little to the migration to metropolitan regions (Table 4.4).

The 'landscape of preferences' for different socio-economic groups is a reflection of in- and out-migration to different regions. These complex flows add up to the net migration patterns for the entire labour force (Figure 4.20). This landscape tends to become very clear during periods when there is demand for labour everywhere and when the housing market functions satisfactory. The second half of the 1980s was a unique period for studying this phenomenon. Factors other than sheer economics and the search for an income were important motives for migration. However, the overheated housing market in bigger cities to some extent hampered the possibility of realising this.

Let us also look at this pattern at the level of the 10 categories of LMAs. Both the total population and the category "young well educated persons" are largely

Table 4.4. Net losses of persons with a higher education to Stockholm 1986-89. Source: Axelsson et al, 1994.

Category of labour market	Age group 16-34	35-50	51-64	Total	(%)
Regional service centres	-1802	221	71	-1510	29
One company dominated	-202	-115	-30	-347	7
Small scale industry	-2037	214	35	-1788	34
Large scale industry	-423	-66	-62	-551	11
Sparsely populated with a dominant regional centre	-350	10	-28	-368	7
Small/medium sized and differentiated	-481	11	20	-450	9
Rural	-168	-46	-16	-230	4
Total	-5463	229	-10	-5244	100

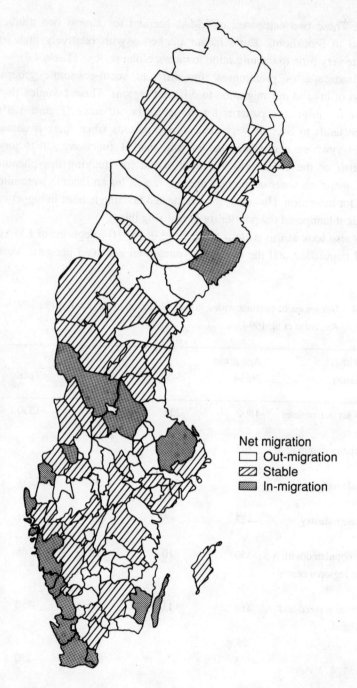

Fig. 4.20. Net migration between local labour markets 1986-89. Age group 16-64 years.
Source: Axelsson et al, 1994.

Table 4.5. Categories of LMAs ranked after migration preferences (quotient in-/out-migration) 1986-89. All people between 16-64 years of age and young persons with a university degree. Source: Axelsson et al, 1994.

Rank among young well educated persons	Rank among all between 16-64 years of age	Category of LMAs	Quotient In-/out-migrants among young well educated persons	All between 16-64 years of age
1	3	Stockholm	1.76	1.07
2	2	Gothenburg	1.36	1.11
3	1	Malmö	1.02	1.19
4	5	Sparsely populated with a dominant regional centre	0.95	1.02
5	4	Regional service centres	0.92	1.02
6	9	Large scale industry	0.79	0.85
7	7	Small/medium sized and differentiated	0.72	0.88
8	6	Rural	0.67	0.94
9	8	Small scale industry	0.64	0.86
10	10	One company dominated	0.53	0.71

"ranking" the 10 categories of LMAs in a similar order, measured as the recorded relation between in-migrants and out-migrants over a period of four years (Table 4.5). For young and well educated persons, Stockholm plays a dominant role with 76 per cent more in-migrants than out-migrants. The young persons' geographical preferences have a more biased and dual character than the total working population's preferences - e.g. small industrial labour markets attract only little more than half as many in-migrants as they lose out-migrants.

Finally in this section, we will illustrate changes in preferences. First, we find that a higher proportion of more highly educated people in the labour force leads to fewer gaining regions and a larger number of losers. At the same time, there are shifts of preferences which to some extent are age specific which also reinforce concentration. Figures 4.21 and 4.22 show two extremes of the settlement system - Stockholm and the rural labour markets.

Almost irrespective of the education level of the labour force, increasing age means a less marked willingness to migrate to Stockholm. Correspondingly,

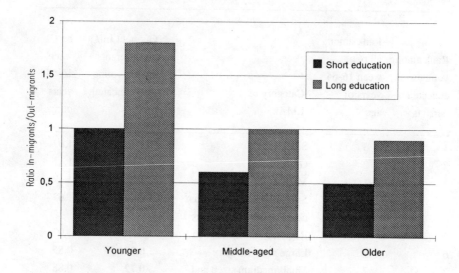

Fig. 4.21. Quotient in-/out-migration to Stockholm region for individuals with and without long education. Three age groups. Source: Axelsson et al, 1994.

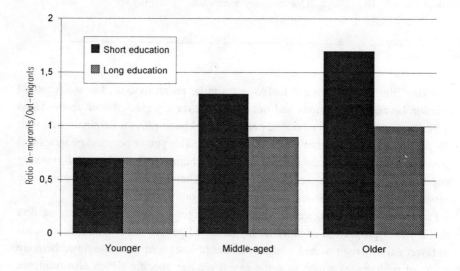

Fig. 4.22. Quotient in-/out-migration to rural labour markets for individuals with and without long education. Three age groups. Source: Axelsson et al, 1994.

increasing age leads to a preference to migrate to rural regions. The patterned preferences for Stockholm are similar to those for Gothenburg and Malmö. The rural pattern is similar in regions with a diversified branch structure.

4.6 Conclusions on Mobility in Sweden

- The geographical extension of local labour markets has led to increased commuting and reduced long distance migration.
- Rural labour markets, regions dominated by large scale industry and one company towns, all had negative net migration for all categories of labour.
- A constantly positive net migration is not found in any category of local labour markets. Southwestern Sweden is the region with the most positive net migration.
- In general, young people and people with a higher education have a strong rate of in-migration to big cities. This is probably a reflection of the attraction of big city life as such and due to the employment options available during the 1980s. Even if the present recession limits the demand for labour, it is probable that the big cities' attraction to more highly educated workers will remain strong.
- Persons without a higher education do not form a homogeneous group in terms of migration. There is a variation according to age, gender and economic sector. Females in public services preferably migrate to medium sized and big labour markets. Males in industry are less mobile and preferably migrate to regional centres and medium-sized labour markets.
- In general, families with children migrate away from big cities and prefer small towns.
- Local labour markets perform different functions for different migrants. Small local labour markets sometimes function as a place for the retirement of elderly people. Some families with children choose small towns since they can provide attractive housing facilities. The primary asset of big labour markets is that they usually can provide a diversified career for those with a higher education.

5 REGIONS AND CONTEXTS

5.1 Comparing Two Contexts

Current changes in Sweden seem to converge into a situation reminiscent of the US; less public intervention in planning, liberalisation of markets and increasing individual mobility. Thus we will present a comparative analysis of the two countries, starting from the following general observations.

The expansion of the welfare state in Sweden proved to have the (initially unintended) consequence of contributing to a relatively balanced regional development in terms of employment and services. The consensus of a concept of equality is important to an understanding of the homogeneous development of public service sectors in each region. However, the levelling out of welfare between regions basically has not changed the centre - periphery relations within the urban system. Already the range in size, from 2 000 to 950 000 inhabitants, of functional labour market areas suggests a persistent polarity (compare chap. 4.3).

From the beginning and throughout the evolution process, the entire welfare model has had a centralised profile, although implementation has been decentralised to municipalities. Municipalities eventually have assumed larger responsibilities.

In the present financial crisis of the public sector, current and proposed reductions of public responsibilities seem to have the most negative impact on peripheral regions. These regions are more dependent on public service jobs and transfers, while the private sector is less developed. European integration continues to proceed and initially seems to have the most positive impact on metropolitan regions.

These processes are transforming the previously well balanced regional development in Sweden into a regional pattern with greater diversity. Conformity is to some extent replaced by equivalency, allowing for increasing importance and visibility of the comparative advantages and local preferences and initiatives. To a certain degree, this seems to be in line with the general trend in Europe in the emergence of more independent regions.

5.2 Socio-Economic Regions in Two Countries

This chapter describes and compares the emergence of spatial patterns of socio-economic conditions in central Sweden and the state of North Carolina, USA. The description is based on a statistical analysis, including a clustering of administrative units, into homogenous regions, using official data on social and economic conditions. The aim is to explain differences in the development of the urban-rural continuum in socio-economic terms[1].

Information is based on county level data in North Carolina and on municipality level data in Sweden. These units are chosen for their comparable aerial size. Counties in the US and municipalities in Sweden are the smallest comparable administrative units containing complete data for both study areas. This study uses the 100 counties which constitute the state of North Carolina and 131 municipalities in central and northern Sweden. The latter are aggregated into 108 data units.

Enlarging on the traditional use of per capita income in the literature as a useful and often reliable indicator of development in technologically advanced countries (Chang, Gade and Jones, 1991), a data set of sixteen independent variables was collected for the Swedish study area. Data were used for three target years: 1970, 1980 and 1990. The variables are listed in Table 5.1. These variables cover a variety of aspects such as demographics, urbanization, education, labour structure, income and commuting. Moreover, the Swedish data compare to the data used to analyse North Carolina.

Cluster analyses of data for the period of 1970, 1980 and 1990 reveal changes in the socio-economic continuum within these two decades. Differences and changes can point to ways to influence the slope of development in an area.

Sweden has three major urban regions - the metropolitan regions of Stockholm, Gothenburg and Malmö. However, only Stockholm with more than one million inhabitants, evidences typical metropolitan characteristics. Most of these regions have shown continuous population growth for several decades, mainly due to in-migration. There is a concentration of private services, both business and personal services, to these regions, especially Stockholm. A large proportion of foreign

---

[1] This section is largely abstracted from Jones (1991).

Table 5.1. Independent variables.

Distance to next higher order city	Population density
% population in agriculture/forestry	Median income
% population in manufacturing	Youth dependency
% population in private services	Elderly dependency
% population in public services	Day to night population ratio
% housing which is overcrowded	Total population change
% population over 25 who have education past secondary school	% population over 25 who have completed secondary school
Unemployment rate	Per capita government transfer

investment in Sweden is directed to Stockholm. Also, most immigrants concentrate in the capital region. During the economic upswing of the 1980s, labour and housing markets became overheated, especially in Stockholm, which resulted in increasing price levels. It is estimated that in the 1980s, the price for real estate for residential properties increased 55 per cent more in Stockholm than in nonmetropolitan regions. However, food prices were reported to be 6 per cent higher in Stockholm than for instance in Gothenburg.

Over the past twenty years, Sweden's urban core municipalities have remained fairly uniform (Figure 5.1). Metropolitan Stockholm is clearly the prime growth centre. On the other hand, central Sweden contains several secondary core regions. Within the part of Sweden examined here an axis of urban centres stretches from Stockholm to the west. A few other core regions are located in the interior, most being administrative centres. To the north, core regions generally hug the Baltic coast. Among the core regions as a whole, there has only been one addition to this group in the last two decades: a manufacturing municipality with industrial activities associated with the military. This municipality registered as part of the intermediate regions in 1970 and 1980.

The vital intermediate regions show considerable change over the last twenty years. In 1970, these regions consist mainly of areas adjacent to core regions in

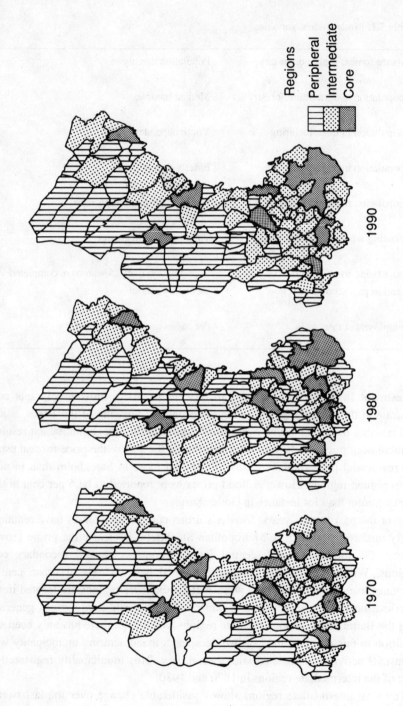

Fig. 5.1. Socio-economic regions in Sweden 1970, 1980 and 1990. Source: Jones, 1991.

the southern portion of the study area. In the northern part of the country, the intermediate regions appear in tight clusters around urban cores by the coast. By 1980 we see a considerable change in Sweden's intermediate regions. The recession of this period, in association with sectoral changes in Sweden's labour structure, caused the heavily-industrialized municipalities to register statistically as peripheries. High out-migration at that time is characteristic of these areas. These municipalities have relatively high unemployment rates in 1980 as well. A six category analysis of the same area in 1990 reveals that these areas are at the lower end of the intermediate regions portion of the continuum. Thus, when recession and economic hardships hit the important single industry emphasis of these areas, their quality of life actually slipped into the upper peripheral end of the continuum.

Sweden's northern Baltic coast is experiencing a gradual increase in quality of life indicators. Indeed, the initial small clusters of intermediate regions around urban cores have expanded to include other adjacent municipalities, which earlier ranked as peripheries.

Altogether, the intermediate regions in Sweden have witnessed fluctuations in several municipalities' quality of life characteristics as they register as intermediate regions and then decline to levels more appropriate for peripheral regions. On the other hand, other municipalities have shown clear growth in quality of life where they have maintained or gained in socio-economic development. Thus, the intermediate regions in Sweden clearly evidence fluctuations both up and down the continuum.

Rural and peripheral areas undoubtedly are much more recognised in Swedish regional policy than are intermediate regions. Statistics and surveys also show that these partly urbanized regions are well equipped with respect to basic services, both in terms of personal and of business services. Labour markets in intermediate regions are generally small but are able to offer employment at almost the same level as in larger regions.

As with the intermediate regions, peripheral areas have also undergone change in the last twenty years. In the 1970s when economic difficulties imposed hardships on the entire continuum, the more developed or subsidised municipalities approached the level of development characteristic of the lesser developed intermediate regions. As strength returned to the economy in the 1980s, municipalities that had registered as part of the intermediate regions dropped to levels of development more characteristic with that of marginal areas. Many of the more remote peripheries have steadily ranked as peripheral in development.

Explicit regional policy in Sweden is oriented largely towards the sparsely populated northwestern interior and to rural areas. Present rural policy is founded on an established idea regarding the uniqueness of rural areas which motivates special instruments: resources are geographically spread and bounded; the economy and the labour market is considered to be strongly tied to the exploitation of natural resources, even if to a large extent the profits of processing accrue to urban centres. Rural firms within manufacturing industry are operating mainly at late stages of the production cycle (Johansson and Karlsson, 1990). Furthermore, it is commonly held that development strategy has to be built upon basic resources and local firms; the ownership structure of the land is fairly fixed; the unique social and cultural systems are considered important to maintain and protect. This perspective guides most of the instruments used in rural policy, such as incentives to small enterprises and income support to small farms. Some of these instruments were introduced as early as the 1940s, but most were introduced from the 1970s on.

However, recent research reveals that most rural areas diverge significantly from this "image" when it comes to their economic and social structure - their functions have changed. In the more urbanized rural areas, public "production of welfare" - i. e. education, health care, social services - now accounts for most of the jobs in the labour market for younger as well as older women. Agriculture and forestry accounts for 30 per cent of the labour market for middle aged men, but less than 10 per cent for younger men who entered the labour market in the 1980s (Johannisson, et al, 1989). Altogether, changes in Sweden's socio-economic continuum under strong government initiatives towards regional development show that areas close to the juncture between socio-economic regions do fluctuate in their levels of development. Although core areas appear more fixed, the intermediate regions are more dynamic with changes (both up and down) in their relative levels of development. These phenomena indicate that the intermediate regions are vital in regional planning and provide the base for socio-economic strengthening from which more marginal areas can be affected.

Although similar in size and population, North Carolina and Sweden differ in the pattern of their socio-economic continua. To begin, North Carolina has three concentrations of primary Core Regions in an arc across the Piedmont area (Figure 5.2). In the Swedish case there is only one primary core, i e Stockholm. Secondary core areas intermesh across the western Piedmont area of North Carolina. These secondary cores link together the dispersed primary cores. Sweden's system of secondary urban cores are dispersed and only approach one similar linkage, i e west of Stockholm. Whereas a third type of urban centre, Urban Centre 2 exists in North Carolina, this tertiary core is not readily identifiable in Sweden. In North Carolina, tertiary cores represent counties where public investment outweighs private investment. For example, these counties are areas which house large publicly funded universities or military bases. In Sweden, however, large scale transfers of public funds to various areas means that public

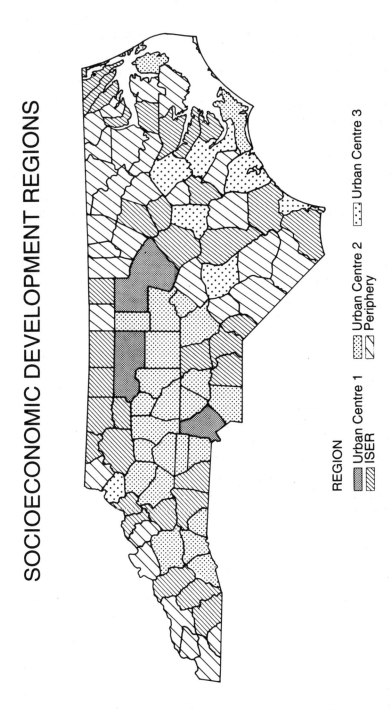

Fig. 5.2. Socio-economic regions in North Carolina. Source: Jones, 1991.

investment is funnelled throughout the entire continuum on a scale unknown in the United States. Although the cluster analysis groups Swedish urban centres into a single category of secondary cores, tertiary core influences can be seen in Sweden.

Both the Swedish study area and North Carolina have almost identical per cent of their populations living in core regions. On the other hand, the intermediate regions contain a considerably greater portion of the Swedish population than their North Carolina counterparts. Likewise, the climatically more hostile Swedish marginal areas contain almost half the percentage of population which lives in the peripheral regions of North Carolina (Table 5.2).

Again, marginal areas in Sweden are largely a product of a distance decay function. Although the Swedish figures reveal the increased importance of the intermediate regions in planning for the Swedish population, they also show the difficulty for Swedish planners in maintaining services in peripheral areas with their low population thresholds and their greater distance from areas with more self-sustaining growth potential. In summary, the Swedish data indicate the importance of the intermediate socio-economic regions as a link between core and peripheral regions. In fact, considering that over half of central and northern Sweden falls within the intermediate regions, its utility is of even greater importance.

As factor analysis reveals, education and job opportunities (other than agriculture) are important in raising income levels and other quality of life indicators. Moreover, a comparison of socio-economic patterns over a twenty year period shows that national, state-wide or international economic hardships can depress the entire socio-economic continuum, pushing lesser-developed segments of the intermediate regions into development gradations usually experienced only in marginal areas. Economic booms may have an elevating effect on the entire continuum as well, but this study does not examine this issue. The Swedish data indicate that the existence of large intermediate regions also aids in the equalisation of incomes in a region. The lack of urban cores to serve as growth

Table 5.2. Comparison of populations by types of region. Swedish study area 1989. North Carolina 1980 Census. Per cent of total regional population.

Study region	Core	Intermediate	Periphery	Total
Sweden	64.6	28.3	7.1	100
North Carolina	61.6	21.2	14.2	100

centres in northern Sweden, however, limits access of peripheral areas to urban services and wages via commuting. North Carolina's poorer coastal plains counties are in a similar situation. Therefore, enhancement through investment in intermediate centres near marginal areas is crucial in alleviating this lack of access to amenities and jobs. Moreover, efficient linkages must be supplied tying core regions to intermediate regions to maximize flow of commuters, services and goods. In conclusion, in many cases the intermediate regions serve as a link and potential development site in both North Carolina and Sweden.

This empirically-based analysis shows that the Socio-economic Continuum model of gradations (see Figure 1.1) in development is applicable for cross-cultural studies of development issues. In North Carolina, the continuum is largely the product of a free market system where education and skill are the key factors influencing the quality of life levels in the different counties. In the Swedish context, however, access is the main factor in defining the continuum. Under Sweden's highly developed social welfare system, large subsidies for regional development programs and income equalisation schemes have lessened the severity of uneven socio-economic development. As a result, the slope of the socio-economic continuum is not nearly as steeply inclined as in North Carolina. Quality of life conditions in general are higher in the Swedish study area than in North Carolina.

The gentleness of the slope in Sweden allows for the intermediate regions to extend further into otherwise marginal areas at a relatively high level of development. Many municipalities in these intermediate regions on the lower gradations of development, however, are dependent on one or two major industries. Recessions or sector changes in employment can upset the balance of these municipalities and cause conditions more characteristic of peripheral regions to appear. Strengthening these economies by diversification of their economic base would greatly help in stabilizing quality of life conditions in these areas.

Sweden, though lacking the closely interconnected network of primary and secondary urban cores found in North Carolina's Piedmont area, has a greater proportion of its population living in the intermediate regions. This situation holds promise in that the population thresholds and infrastructure needed to build conduit linkages into and through the intermediate regions to connect core and peripheral regions are already present in many cases. Governmental stimulus on transportation and development corridors directed at marginal areas appear to be beneficial approaches to creating greater access to jobs and services for people in marginal areas.

In Sweden as in North Carolina, intensive, large-scale public investment in universities, military bases and other long-term institutions can provide a major boost to an area's socio-economic conditions. This type of infrastructure investment in marginal or intermediate regions areas with proper transportation links to urban cores might serve as a method of outreach to peripheries in crisis. This approach is, however, expensive and requires the development of a parallel service economy for these institutions to function well.

Finally, the example of income equalisation in Sweden illustrates that narrowing the gaps between rich and poor regions is not futile. Sweden's society, with its high taxes, has of course paid a price for this regional smoothing of inequalities.

5.3 Sweden in the European Union

As Sweden becomes member of the European Union (EU), we may face a situation where the economic and institutional framework of Sweden comes closer to the present situation of a single state within the US. The main rationale and vision which inspires economic integration in Europe is to establish an effective common market for production which is sustainable and fully competitive globally. Certainly, the outcome of this global competition will both depend on, and have consequences for, development in the regions of Europe. Global and regional development are becoming more interrelated.

However, there is great uncertainty involved in the current processes of European integration; the uncertainty of the geographical and economic size of the EU in the future and the prospect of expanded integration not just of present EFTA countries but also some eastern European nations. This uncertainty is evoked by the difficult process of putting the Maastricht Treaty into practice in each country and involves the fate of the nation state as well as the function of different regions. In a social and economic project of the size of the new Europe, it is evident that there is great uncertainty about the whole spatial distribution of population and production and the role of existing and new infrastructure.

The initial aim of the EU was also to outline instruments for promoting effective political decision-making within the Community by strengthening the economic and political role of the Commission. The re-establishment of the original (from the 1950s) ideal of the Common Market through free trade and deep-rooted integration was expected to replace the "withdrawal into national egoism" that was perceived to be widespread in Europe during the 1980s.

The content of the original vision of the single European market is probably best understood by re-examining the most important measures that were suggested and implemented; the removal of border controls, the opening up of public procurement, the harmonisation of technical standards, the harmonisation of indirect taxes and the liberalisation of financing services. The vision is of a Europe which allows for strong and unfettered competition among producers in all member states, where the real factor and transport costs of each site will be transparent and decisive for success. This vision may, in a somewhat caricatured form, be reminiscent of the idea of 'survival of the fittest', but at the same time the very idea of a Community places emphasis on solidarity between all regions; favoured and less-favoured.

Some researchers conclude that the single economic market will lead to a worsening of regional problems in Europe (e.g. Boltho, 1989). In short, the argument is that peripheral regions, by definition, have locational disadvantage due to relatively increased transportation costs and also have difficulties benefiting from large scale economy. The suggested reforms of the Structural Funds are, by many observers, not expected to be sufficient to compensate for disparities between cores and peripheries.

On the other hand, some officials give support to the idea that transport costs are becoming less important in the location decisions of manufacturing industries, partly as a consequence of the constantly higher value per weight in industrial products. Combined with notably strong developments in telecommunications, there are arguments supporting the vision of a prosperous future for peripheral regions in Europe also (Delors, 1989).

During recent years has also been observed a transformation pattern in Europe, which has been labelled *regional inversion* (Suarez-Villa and Cuadrado Roura, 1993). This dynamic development is appearing in some peripheral, often border, regions due to changes of the technological, political and economic conditions.

Some authors argue that the emerging integration will lead to radical changes and adjustments in the institutional framework (Veggeland, 1993). Power is expected to move higher up within the system and favour strongly centralised governmental bodies in a European federation. This is likely to lead to international conflicts and institutional crises will threaten its foundation. History has shown that successful federations are characterized by inhabitants possessing common background in culture, language and values. According to these authors, the conclusion is that the concept of a Europe consisting of regions is most likely to succeed. Regionalism implies that power and initiative is decentralised to regional institutions, following the principle of subsidiarity. In this vision, future European integration is envisaged as a regional mosaic pattern, consisting of dynamic and stagnating regions, which replaces the former uniform concentration of economic growth into national core areas. There is less determinism than in the earlier, industrial society where a core - periphery dichotomy was ubiquitous. The mosaic model implies that opportunities are more open and that local conditions are of decisive importance. The list of important local conditions contains political conditions, infrastructure and physical planning, supply of qualified staff, cultural conditions and modes of life, factor prices, and density of population.

In this process, regional initiatives and central EU decisions are becoming more important than national policy making. Urban regions are competing for employment, income and skills. Both local assets and interregional infrastructure are becoming important in this competition. Skills and technology become more important in the new spatial division of labour.

6 POLICY AND PLANNING PERSPECTIVES

6.1 Introduction

The characteristics of the Swedish socio-economic continuum is, to a considerable extent, influenced by planning and policy-making. In this chapter we describe long-term and current changes in macro-economic, political and planning conditions. As a result of the emergence of more individual rather than collectivistic values, private rather than public initiatives and increasing periodic mobility for many individuals, a social change is taking place in many regions, rural as well as urban and central as well as peripheral. The important administrative and planning region of the municipality as well as the functional region of the local labour market change their signification. The functions of different regions are becoming more mixed at the same time as each region is trying to develop a profile in order to meet increasing international competition.

As in previous chapters, policy and planning in the United States is used as a reference for the on-going reorientation in Sweden.

6.2 Sweden 1930 - 1990: The Central Role of the Welfare State

The evolution of the Swedish welfare state began in the 1930s under leadership of the dominant Social Democratic party. An informal alliance developed among organized labour, the state and major private capital. Equalisation of living conditions in different social strata was one over-riding goal, popularly expressed by the metaphor of "The People's Home", already formulated in the 1920s. The main principles were large scale intervention by central government in the economy, centralism and uniformity. High levels of international competitiveness within industry resulted in expanding economic resources in the national society.

This created financial possibilities - first largely through taxation on labour, later also on consumption - for an expansion of social responsibilities in the public sector. Production of welfare services became organized within the framework of numerous public sector institutions. The widespread use and legitimacy of the model is clear from the 44 years of unbroken Social Democratic government from the 1930s through the 1970s.

The real take off for the welfare state came during the 1960s when it was underpinned by strong economic growth, later referred to as "The Record Years". Welfare benefits accrued not only to the poorest but also to the middle class. During the 1970s the welfare state created paid employment opportunities for a great share of women; in practice, informal and unpaid female jobs within families were transformed into formal jobs in the social and health care sectors. Private entrepreneurship and competition was largely excluded not only from education, health care, social care and associated sectors, but also from urban and regional transportation, telecommunications, airlines and railroads. Private initiatives in many sectors were restricted and regulated, such as housing. As a general rule, basic public services were offered free of charge. To some extent, Sweden became a model state, as a high private living standard (in spite of high personal taxes) was accompanied by probably the most ambitious (and expensive) public safety net in the world, all with a strong egalitarian profile.

The labour market and economic policy programmes of the 1960s focused on stimulation of further economic growth by supporting the transfer of people from low productivity sectors such as agriculture and forestry to dynamic manufacturing industries. The regional policy programme which was formulated in the mid 1960s concentrated on incentives to industrial location in the less developed regions which were experiencing a net loss of labour (primarily the northwest). The historically successful export orientation of large manufacturing corporate firms was a necessary precondition for steady growth. A dominant feature of the urban system in Sweden was and is still the widespread presence of branch plants and stable subcontractors to the big forestry based companies and other major manufacturing companies, i e Volvo, Saab-Scania, SKF, Ericsson and Asea Brown Bovery.

The direct or indirect dependence of the national as well as the regional economy on a few successful companies and the relative lack of experience of small scale private enterprises may be one explanation behind the widespread consensus about the welfare system. For several generations there has evolved a pattern of patronage - from the side of capital owners and the state - and a trust in collectivism and solidarity within the labour force. The large size of a small group of Swedish-owned companies compared to the national economy contributed to an impression of a harmonious national alliance between the state, private industry and the labour force. However, during the 1970s the Swedish economy faced problems of stagflation, unemployment and high militancy from labour unions which put pressure on the government to institute a Keynesian defence of the welfare goals. Maintaining a low open unemployment rate became one of the most important strategies. Some of the traditional Swedish industries such as the iron and steel industry and shipyards experienced strong international competition and suffered from decreasing market shares. It became evident that the cost of production in Sweden had increased above the level of international competitiveness, partly due to the demand of resources by the public sector but also because labour unions had negotiated successfully for rapidly increasing wages. The unions had agreed on a principle of solidarity which was interpreted as the same wage for the same job in each industrial branch regardless of the location or the firms' ability to pay. This put pressure on less productive units in peripheral regions. However, as regional policy had simultaneously turned towards a distributional function, several means were available to compensate for closures or other dramatic reductions of the needed labour force. In fact, regional equalisation became superior to stimulation of economic growth.

Further examination of the details of Swedish regional policy allows us to identify four phases since the middle of the 1960s. Phase 1 (1965-72) was concentrated on modernisation of lagging rural and urban areas. Efforts were focused on support to industrial firms and basic services. Investments were also made in transportation facilities and improvement in the public transport systems. Phase II (1972-76) was characterized by a municipal reform which was accompanied by a rationalistic plan for the further development of the centres of the municipalities and relocation of several governmental authorities from Stockholm to medium-sized cities. Phase III (1976-85) meant a considerable shift from the principle of a ruling central government to stress on the mobilisation of local resources. During this period the rationalistic plan for the future regional structure was discontinued. In Phase IV (1985-) emphasis is placed on improvement of infrastructure for rapid transportation and communication as well as the further development of professional and technical competence and cultural institutions in lagging regions. The major new measures also include decentralisation of higher education, and special education and technology diffusion programmes.

A consequence of these changes in policy and planning is that ISERs have become (unintentionally) more visible, in a positive sense, in the regional and national context. The different generations of modernisation programmes have both upgraded these areas with urban qualities and promoted the creation of linkages not only to the nearby urban region but to a multiplicity of urban regions.

6.3 The Intermediate Role of Municipal Planning

Sweden's first local government legislation since the medieval times was passed in the 1860s. Prior to that parishes had provided both ecclesiastical and social services. The major task for the new municipalities was to provide poor relief. Of the 2 500 municipalities, more than 95 per cent were rural. In fact, only 11 per cent of the population was living in urban areas (defined as places with more than 200 inhabitants). Even in the 1920s, the corresponding figure was not more than 35 per cent. Compared to most countries in Europe, Sweden was urbanized rather late.

As urbanization proceeded, municipalities became more involved in the social sector. In 1917, municipalities were instructed to take active part in the improvement of urban housing standards. In the 1920s and 1930s, further municipal obligations were introduced both in the school system and in social welfare. The increase in local production of services led to the fact that by the 1930s, municipal production exceeded 10 per cent of GDP. In general, the social problems involved in the on-going urbanization were the most important motive for the expansion of local government (Häggroth, 1993). Rapid expansion and the professionalization of the services and the administration provided by local government demanded large-scale solutions in order to be cost-efficient. It became obvious that small rural municipalities especially were not competent to handle these large projects. To strengthen the local ability to fulfil egalitarian principles a municipal reform, focused on a considerable reduction of the number of municipalities, took place in the early 1970s. Central place theory formed the theoretical background for this reshaping and reinforcement of the lowest administrative level. It was recommended explicitly that each municipality should be a functional unit - in terms of labour, housing and service markets - embracing both less-developed rural regions and urban centres. The number of municipalities, which had, by the 1950s, already been reduced to 1 000, was further reduced to 284. Ideally, municipalities became functional units adjusted to host responsibilities for the expanding set of welfare services. In practice, the preconditions still varied, since the size of the new municipalities ranged from approximately 3 000 to 700 000 inhabitants, with an average less than 20 000.

Long-term economic planning by the central government became a tool in the process of implementing new social reforms. The central government also decided that municipalities should prepare long-term investment plans. Although municipalities were raising local income taxes, they were, to a large extent, depending on the national redistribution system of tax resources via the state budget. Municipalities with decreasing and ageing population and with a weak taxation base (because of a dominance of less productive industries) were almost fully compensated by means of these transfers. However, the use of the municipalities' financial resources was restricted by detailed rules concerning the standard of schools, social welfare, public housing etc. In retrospect, at least, the general impression is that local governments all over the country, even those with domination of conservatives, liberals and farmers, developed into satellites of the central government. Our conclusion is that the growth of local public services became the strongest instrument for a spatially equal distribution of employment and welfare in the period between 1960 and (at least) 1990. Compared to specific regional policy (i.e. incentives to industrial firms located in, or relocating to the northwestern aid area), the spatial impact of the distributive and compensatory function of public sector has been much stronger.

6.4   Approaching a Bottom-Up Planning Model

For many centuries, a top-down planning and implementation system has dominated in Sweden. Subsequently, however, changes have occurred. Initially it affected physical planning. In 1972 a national physical plan was presented by the central government. This plan had a strong impact on municipal physical planning in terms of delimiting environmentally sensitive areas. It provided strict guidelines for land and water use and places of historical importance. Municipalities elaborated physical plans for land use, suggested spatial organisation of social and physical infrastructure and sometimes also presented a strategy for implementation and solutions to conflicting land use interests. It was recommended that municipalities follow a special planning procedure which corresponded to national policy objectives.

Since the new legislation of 1987, both in theory and practice, Swedish physical planning is turning from a top-down to a bottom-up approach. Each municipality is responsible for a comprehensive plan for the use of the whole territory, land as well as water. The plan is to be decided by the municipalities themselves, not as before by the state. County government is able to temporarily veto a plan only if it does not meet important national interests, if it is not sufficiently coordinated with neighbouring plans and/or if it does not provide the necessary health and safety conditions to the citizens. As Bylund (1992, 1994) points out, physical planning in Sweden has certainly turned into a bottom-up procedure, although the size of local administration and the complexity of municipal democracy has until now largely excluded grass-root participation in the planning process. Big municipalities usually have competence and resources to develop efficient and sophisticated physical plans according to the planning law, but usually with little grass-root involvement. Small municipalities often have a larger public involvement, although the plans usually have to be developed by consultants. In spite of the right to veto the plan if neighbouring municipalities' problems are not sufficiently respected, it is very common for local comprehensive plans to neglect inter-municipal problems.

Almost all local government activities used to be financed and produced under municipal auspices. Competition from private activities was almost absent. Over the past few years, this image of local governments is less valid. Public monopolies are being relaxed and competition is more common. According to recent legislation, all local government activities not exercising authority can be conducted on a contractual basis (Häggroth, 1993). As a result, an increasing number of municipal companies are working in the technical sphere. Another alternative is to transfer production to private contractors. Contracting is more widespread in smaller municipalities. One of the latest innovations is private firms assuming parts of social service functions, e g service flats for elderly people. Co-operative firms are becoming a new way to produce services, mainly in the child care sector. Finally, municipalities are transferring operational responsibility to voluntary associations.

One rationale for this is the need for a less sectorial division of local planning. It is generally recognised that the preconditions for a territorial overview and cross-sector planning are better in medium sized than in either metropolitan or rural municipalities. The favourable prerequisites for territorial considerations in intermediate municipalities is the combination of rather weak sector administrations and relatively strong local competence. In metropolitan municipalities the efficiency of each sector, involving scale economy, becomes key goals promoted by strong and professional administration of each sector. In rural municipalities, the sectoral division of the rules and the resources provided by the state becomes difficult to overcome. Negative or ineffective threshold effects appear.

6.5 Alternative Planning Modes

From the perspective of assessing the relevance of planning in influencing the intermediate regions there are two dimensions that need elaboration. These are rationalist planning or negotiational planning, and centralisation - decentralisation.

The first dimension places emphasis on how the planning procedure is carried out, the second places emphasis on hierarchical aspects of policy-making and planning.

In this framework, Friedmann (1987) presents four conceptual modes of planning: (a) policy analysis (rationalist planning); (b) social reform (planning for social change through incrementalism); (c) social learning (planning as learning through participation: (d) social mobilisation (autonomous action-oriented planning). In discussing how the concept of regional planning has changed in character in Sweden, and the other Nordic countries, since the 1950s, Veggeland (1990) argues that these four conceptual modes of planning have appeared complementary to each other. Successful planning results have often been based on duality relationships among rationalist, participatory and negotiational planning approaches. Friedmann labels this 'transactive planning'. The decentralisation trend during the last decade has promoted flexibility in mixing different planning approaches, and with flexible sets of actors. Despite variations among planning modes used during each time period, the main stream in this transformation process follows the circular pattern in Figure 6.1.

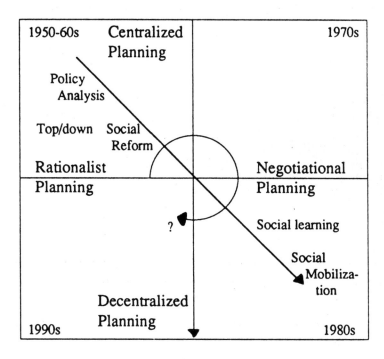

Fig. 6.1.   The circular motion of planning. Source: Veggeland, 1991.

6.6 Policy and Planning in the United States

The case of the United States presents an interesting reference.[1] There we have few state-wide examples of implemented comprehensive regional planning strategies. Among the Friedmann conceptual modes only "policy analysis" might be recognised as a component of the planning effort. In this way the regional impact of relatively unfettered capitalism may be diagramed as in Figure 6.2.

Here we see the highly generalised socio-economic continuum, cost and revenue effects of open market, and free factor mobility conditions. Revenues have an inverse relationship with distance from the core (principal market) since the cost of moving products and services from point of origin to market is largely a function of distance. On the other hand, the costs of assembling and implementing the factors of production are a function of traditionally lower factor costs in core areas (capital, skilled labour, technology, infrastructure). Thus these costs will increase with increasing distance from the core, or from any adequately scaled urban centre. Locations in the shaded area, where revenues exceed costs, are competitive and will tend to attract investments on their own accord. Given this situation, investors have generally chosen to locate in or near the metropolitan

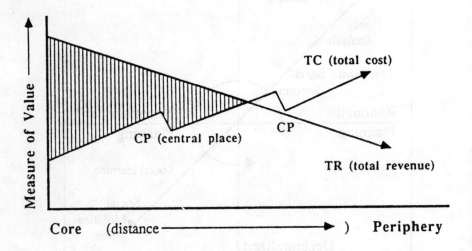

Fig. 6.2. Production cost and revenue relationships along the urban-rural continuum (after Stöhr, 1974).

---

1 This section is largely abstracted from Gade, Persson and Wiberg (1992).

core. This speaks in favour of a continuing spatial differentiation in production demand, or access to markets. In the United States, these locational aspects clearly are reflected also in social behaviour, with similar spatial parameters established for population settlement patterns and the gradation in quality of life conditions. Thus the hypothesized total revenue curve in Figure 6.2 bears a striking resemblance in its distance decay function to the quality of life curve in Figure 1.1. Where the two curves meet, we may reasonably suggest, is the beginning of the periphery. An exception from this general pattern is that some ISERs have benefited from the tendency of individual states to place institutions of higher education (regional universities as well as community/technical colleges), large prisons, mental health hospitals and other public facilities in predominantly rural areas.

It is relevant here to illustrate our perception of existing differences in total costs and returns between free market and welfare state conditions. As indicated in our introduction, we view the slope of the free market economy quality of life dropping more precipitously with increasing distance away from the metropolitan core than is the case for the welfare state economy. Figure 6.3 contributes to this perception by demonstrating the greater slope of the free market cost curve, being unaffected by the regional equalisation policies implemented by the social welfare state. The latter situation provides a less angular and softer slope to the cost curve.

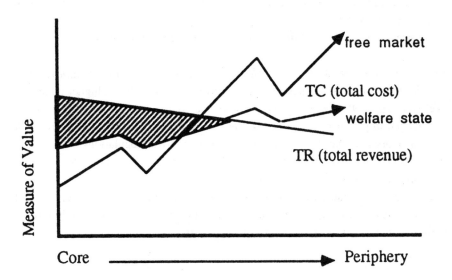

Fig. 6.3. Production cost and revenue relations as a function of welfarist or capitalist development policy (after Stöhr, 1974).

In the United States there are examples of planning with a strong bottom-up approach. For example, Reiman (1992) presents how a strategic initiative for the economic development of a marginal region in North Carolina was undertaken in the early 1990s. A broad representation of local people was invited to participate in the work of identifying critical issues and securing participation in the strategic initiative. From these discussions emerged six task forces: (a) education, literacy, and workforce preparedness; (b) effective government; (c) image improvement; (d) industrial infrastructure; (e) industrial incentives; and (f) economic diversification. In the task forces a great number of practical ideas were suggested. Below are listed some examples:

- offer of free environmental studies, free new sites, and free utility access and upgrades to industries;
- lowered property taxes;
- lowered utility rates;
- improvement of access to rapid long distance transportation infrastructure;
- development of an industrial park;
- incubator for small businesses;
- targeting specific businesses and industries in order to improve the overall mix;
- establishment of appropriate capital reserve funds;
- raise educational levels and better adjustments in general to the demands of modern workplace;
- strengthening of local links between government and business.

Especially worth attention is the emphasis which was put on image improvement in order to attract new businesses, "particularly those that allow a higher quality of life" (ibid., p. 119). It was suggested that outside public relations expertise should be engaged in preparing a detailed strategy to be used by the local government. It was also argued that land development agencies use regulations in order to improve the physical appearance of the region and better control natural environmental impacts of industries.

The suggested strategies are now facing an implementation procedure. Some of the strategies require long term funding arrangements, while others only require enthusiastic, entrepreneurial and co-operative people.

6.7 Sweden in the First Half of the 1990s

General economic conditions in Sweden have changed rapidly during the last few years. Growth of domestic demand which supported the upswing has decreased. At the same time, Swedish producers' market shares are dwindling rapidly because of declining competitiveness. Thus little support for growth can be expected from the external side.

Consequently and according to both external and internal observers, the current Swedish economy is considered to have severe problems (OECD Economic Surveys: Sweden. Paris, 1991). For a long time, the overall growth rate has been lower than in most European countries. Total factor productivity increased by 0.61 per cent per year in Sweden during 1970 - 1985, while Western Germany had an increase of 1.21, the US an increase of 1.66 and Japan an increase of 3.29 per cent per year (Swedish Productivity Delegation, 1991). The balance of foreign trade has been negative for several years. The rate of inflation is high and seems to be caused by structural deficiencies of the economy which lead to distortions of market signals.

Unemployment figures have increased dramatically since 1991. At that time the average level was 3 per cent, including those in special employment programmes. In 1994 the average unemployment is approximately 10 per cent. This is a unique situation during the welfare state period.

Some of the distortions are considered to be related to the Swedish public sector and the way it is financed. The public sector in Sweden is the largest in Europe in relative terms and its influence on private sector behaviour is profound. Taxes represent 55 per cent of the GDP, while the OECD average level is 39 per cent. Public consumption amounts to 27 per cent of the GDP in Sweden. Furthermore agriculture, which is supported well all the way up to the Arctic circle, creates a persistent problem due to surplus production and heavy subsidies. Sweden's assistance to agriculture corresponds to 52 per cent (in 1989) in terms of Producer Subsidy Equivalents. This is approximately 10 per cent units higher than in the EU (Viatte and Cahill, 1991).

An independent analysis shows that these macroeconomic problems are expected to continue for several years, with a slow growth rate, reduction of the state budget and increasing unemployment. On the other hand the Swedish economy is generally thought to have certain strengths (Swedish Government's Financial Plan 1991/92). The industrial sector is developed. In spite of increased unemployment during the last few years, employment is still high from a European perspective and the labour market is considered to be quite efficient.

Incomes are more evenly distributed than in most other countries. The welfare system encompasses everybody and reaches almost everywhere.

Over the 1980s, support had decreased both for further expansion of the public sector and its concomitant increase in taxes. Changes in the international arena, with the fall of neighbouring socialist states in Eastern Europe, contributed to an increased resistance to central planning and state intervention. As shown in Figure 6.4, the proportion of people demanding reduction in the size of the public sector increased from about 35 per cent in 1986 to a peak of 55 per cent in 1990. The election to Parliament in September 1991 reflected these shifting attitudes. Around 1990-91, less than one fifth of the voters objected to a reduction in the public sector. The recession which started at almost the same time, however, had an important impact on public opinion. Three years later support for the public sector was as strong as (or stronger than) it had been in the mid 1980s.

Already by the late 1980s, the Social Democratic government began implementing measures in order to enhance efficiency in general and, in particular, in the public sector. Between 1991 and 1994 Sweden had a centre-right coalition in power in the central government which proclaimed that they would change Swedish society in certain qualitative respects. The recipe was neo-liberal. We identify three basic elements in the program for Swedish economic policy. First, the aim was to take a more active role in the European integration process.

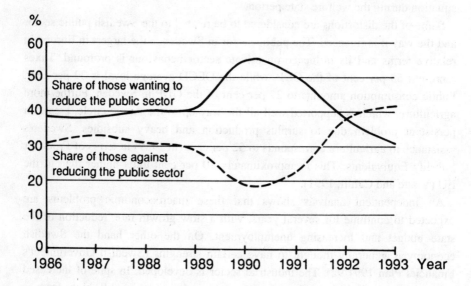

Fig. 6.4. Opinion on the size of the public sector: 1986-1993. Source: Opinion polls reported by S. Holmberg, University of Gothenburg.

The obvious evidence of this strategy was Parliament's decision in 1991 that Sweden should apply for full membership in the EC/EU. One important reason for this sudden policy shift is probably that major Swedish companies, heavily export oriented for a long time, started investing more abroad and increased joint venture agreements, mainly within the EU, during the latter part of the 1980s. In 1989, almost 30 per cent of business investments by Swedish firms occurred abroad. However, the main official reason behind the Swedish application was that the conditions for Swedish security and neutrality policy had changed radically as a consequence of the dissolution of Soviet Union and related political changes in Eastern Europe (The Prime Ministers' Declaration to Parliament, 1991). A second element of the policy was to reduce production costs, mainly by limiting increases in wages, and improve competitiveness in order to avoid unemployment and safeguard basic parts of the welfare system. The third policy element was to reinvigorate growth by governmental and private investments in infrastructure, by job training and promotion of formal qualifications of the labour force.

Below is presented a further examination of changes in policy-making during the first half of the 1990s:

- For transfer programs the major impetus has been to lower compensation levels. Interest subsidies have been cut within housing policy. Sickness benefits have been cut in several ways: compensation levels have been reduced from 100 per cent to a low of 65 per cent and a high of 90 per cent of regular salary. A one day wait before benefits are paid has been introduced. Responsibility for payment of these benefits has been shifted from the state to employers. Altogether this has reduced absenteeism. This is certainly also reinforced by the current labour market situation.
- To cut increasing costs for labour market policy, unemployment insurance now covers only 80 per cent of regular wages. Relief work wages have been reduced by 10 per cent. The pension system has also been changed, among other ways, by reducing the possibility for early retirement.
- Most social services are produced by municipalities and counties. As mentioned, central government had strong influence on the allocation of resources in each municipality. The new trend is toward stronger independence, reinforced by the way resources are now transferred to local authorities, i e largely in the form of lump-sum payments. This leaves more room for local priorities. The municipality of Stockholm, the largest in Sweden, has in many cases been leading the way in introducing new organisations of health care, schools and other services.
- Local priorities, within certain limits, also impact the local income tax rate. There is a tendency for high-income municipalities to keep local taxes low,

while low-income municipalities usually have to finance expensive social programs by increasing the tax level. The rapid shift from an overheated labour and housing market to a reverse situation is causing a number of municipalities to experience financial crises.

The reorganisation of social services in recent years has followed three steps: first there has been a move to try to improve efficiency within the existing organisation, secondly service producers have been allowed or forced to compete, thirdly privatisation has been introduced. Private schools are operating with 85 per cent of their costs covered by the government; as long as they follow the core of the national curriculum they are free to choose their own profile. In 1994 it is estimated that 20 per cent of the services financed by local governments are produced by private firms. Only two years ago the corresponding figure was a scant few per cent. The principle of free access to all social services is being abandoned and fees are being introduced for many services.[2] Altogether, we anticipate seeing increasing differences between municipalities due to local policies regarding service subsidies and accessibility. The general principle still remains, however, that there should be public control of the quality of services and a regulation of fees.

As part of the policy, the traditional planning model which was aimed at creating similarities and equalisation was more or less replaced by a generative bottom-up planning approach promoting differences. Central government policy seemed to focus on increasing international competitiveness by stimulating regional development in dynamic expansion rather than supporting a balance in regional development and maintaining full employment. Municipalities and regions were encouraged to promote and advance urban and regional entrepreneurialism. This means that more responsibilities were given to the local and regional level but also more freedom to develop unique concepts based on public-private partnership and the formation of alliances among different municipalities or regions.

This stress on local and regional mobilisation and entrepreneuralism is a new element in Swedish policy and planning. As discussed earlier the traditional role of local and regional institutions is to act in accordance with central government policy and regulations; they have formed part of strong vertically integrated decision systems. In practice approximately 2/3 of the average budget for a municipality is still used to fulfil regulated service production.

---

[2] The price for consulting a medical doctor has increased from 7 SEK in 1970 to 120-180 SEK in 1994. In fixed prices, this means an increase of between 330 and 500 per cent.

Figure 6.5 illustrates the on-going change in policy and planning in Sweden. The vertical axis represents the gap between centralised planning and informal problem solution in local communities. The horizontal axis represents the gap between a holistic and a selective welfare approach. We argue that Sweden is found in the upper right corner but is gradually shifting towards the lower right corner.

The Social Democrats came back in power after the elections in September 1994. They have declared a policy putting more stress on the traditional egalitarian profile of the Swedish welfare model, although there is yet no signs of a "nostalgic" revival of the previous model as a whole.

At any rate, we can see new ways of thinking being spread in Sweden. The society is becoming less homogenous in several ways. Immigration patterns are a very obvious part of this process. National uniformity is being replaced by more unpredictable social processes. Fragmentation and complexity increases parallel an increase in international contacts and influences. Stabilizing institutions and behavioural patterns are challenged. We can observe a decreasing political influence of organized labour.

The implications of these new conditions is that the government is no longer the strong and dominant planner and organizer of Swedish society. The possibilities of reaching a strong consensus in public planning will decrease. A process has already started moving towards a more market-oriented situation with a plurality

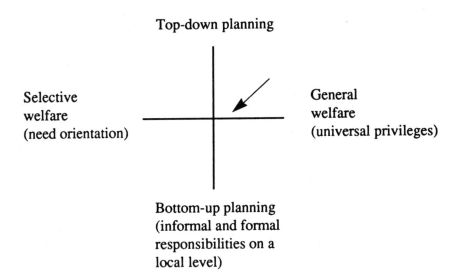

Fig. 6.5.   Trend in reorientation of welfare arrangements in Sweden.

of involved actors - private and public, domestic and foreign. In fact this means a shift from efforts towards national minimisation of differences in quantitative welfare through uniformity to efforts to generate competitiveness and attractivity through heterogeneity based on unique comparative advantages.

7 SWEDEN FACING A NEW MICRO- AND MACROREGIONAL FRAGMENTATION

7.1 Uniformity and Fragmentation

Paradoxically, contemporary Sweden is characterized by both uniformity and fragmentation. Historically, the territory which is now mentioned as Sweden, consisted of relatively independent regions, the 'landscapes'. In early medieval times, i e until the 15th century, each of the 24 'landscapes' had its own legislation. Today and for several hundred years, the 'landscape' has no administrative meaning although the concept is still alive and widely used in common language as a symbol of the macroregional identity. As an 'image' or trade mark, the concept is also sometimes used in regional marketing, e g within the tourist industry. Each of these regions had a more or less distinctive culture. Economic integration was sometimes restricted by trade barriers, but generally practised both among regions and with foreign countries.

Gradually, through the national unification processes in the 16th and 17th centuries and through the political reforms during the last three centuries, Sweden became a national state with a considerable degree of unity and uniformity. Cultural and social conformity was stressed actively in the democratisation process. However, local and regional cultural profiles are still visible and widely recognised. Increasingly, local or microregional profiles including natural assets are promoted in different types of geographical marketing. Economic integration among regions is promoted by investments in infrastructure for transportation and communication and through various measures within the framework of regional policy, industrial policy and labour market policy, including wage policy. Yet, the industrial structure varies considerably among regions. A number of industrial regions depend almost entirely on raw material production processes, while some other regions have been transformed into attractive regions for information and knowledge oriented service industries. The development of the public sector has, however, contributed substantially to a distribution of service jobs to all local labour markets.

There is now a tendency of increasing fragmentation in terms of regional specialisation and local empowerment. By fragmentation, we more precisely mean the spatial impact of:

- an increasing variation and segmentation considering the quality of the jobs available;
- an increasing variation of the quality and accessibility of individual services;
- an emerging socio-economic polarisation.

It appears inevitable that the spatial and social distribution of wealth is becoming more uneven. This new pattern is likely to be repeated at both the micro- and macroregional levels. As this process is characterized by a high level of uncertainty in details and spatial consequences, policy makers and planners have difficulties dealing with it. They can not use ready-built frames of reference and patented strategies from recent decades. Recently similar processes going on, also in other parts of Europe have been highlighted by the European Commission.

"Dramatic contrasts such as those between the centre and the outlying regions are being overtaken by a more complex pattern of territorial organisation.... This diversification of disparities is generating a patchwork in which privileged areas border directly on depressed areas" (Millan, 1993).

Throughout this book, we have tried to explain the processes which have led to the new situation. In this last chapter, we want to summarize and synthesize the outcome of the prevailing processes, and try to envisage the emerging spatial pattern in Sweden. Again, we focus on the future of the intermediate regions rather than metropolitan or rural regions. The purpose of this chapter is also to outline some of the challenges for policymaking which are involved in the changes that we anticipate.

The underlying idea is that micro- and macroregional fragmentation may be perceived as 'new' in terms of its physical and socioeconomic structure and in terms of the problems that emerge from these structures, but at the same time we argue that most of the processes have been present and at work for a substantial period of time. As the processes have been slow their potential for causing a dramatic change has been underestimated. Some of their effects have been hidden by more dominant features or even by political interventions. In this way, our comparative approach between Sweden and the US has proven to be fruitful in revealing some of the underlying processes which are common or different in these two institutional and political frameworks.

7.2 Four Carriers of Change

Here we highlight four general driving forces which all lead to further regional fragmentation: (i) internationalisation and organisation of economic activity, (ii) diversity of life styles, (iii) mobility patterns among people, and (iv) policy formation and practices. It should be clear that all of these changes are supported by - or at least linked to - the transition of the Swedish welfare state into a more market-oriented system. This means that references to corresponding processes of change which we have found in studying the US are implicit throughout this chapter.

Internationalisation implies, among other things, that private enterprises are becoming dependent on exploiting and developing markets, and creating joint ventures with partners located at long distances from head quarters. Production sites will be chosen with little concern for national borders but rather according to local cost structures and accessibility to relevant infrastructure. Technological development, the liberalisation of international trade and the emergence of mass markets are the major factors facilitating this development. To maintain and reinforce competitive strength, all enterprises have to adjust their activities to the new order of production and stronger international competition. New organisational modes, including less hierarchical and more network based management principles, are essential in this development. This kind of organisation is appropriate to achieve flexibility and speed.

At the same time, the microregional locality remains and evolves as an important base for production, especially for knowledge based enterprises. The quality of the physical (natural and built) environment and cultural diversity play important roles in attracting key persons and competent labour in general. This means that local growth can be promoted successfully by efforts to develop specific local infrastructure. In this process it is very likely that success often feeds upon success, i e that attractive sites also attract investments in high-quality infrastructure. But local development does not mean isolation. Hence, it is crucial that the local community is provided with effective links for transportation of people and goods as well as for information through telecommunications to other regions. However, a large number of firms will not be able to liberate themselves from their "home base" even if it has poor qualities in terms of business services and professional labour. Their only way to stay in business will be through price competition, meaning that the wage level has to remain low. At the microregional level, this will contribute to the survival or emergence of a number of small industrial towns with a generally low income level.

The second driving force leading to increased regional fragmentation, is the tendency towards more individualistic life styles. Throughout the long period of industrialisation and the rise of the public sector, the life style of the wage earner was developed and became dominant. This life style developed into a rather homogeneous form characterized by collectivistic and materialistic features. The local community, with its cultural traditions and social networks, was providing an important coherency. The nationwide equalisation of living standards in terms of real income, housing and education was a commonly held value. Among the young generations, we now observe the development of a new life style with more pluralistic characteristics and influenced by global trends, with all their deviations. This image reaches young people in all regions simultaneously through the media and extensive travelling. The growing number of immigrants from other countries with quite different cultures which are spread all over Europe forms another potential for diversification of life styles and development of more heterogeneous local communities. Together with an increasing level of general education and the emergence of knowledge and information based small production units with flexible working conditions for labour, individualistic lifestyles are fostered. This has implications that it will be possible, and preferred, by a growing share of the population to make their own decisions about where to live and work and make these arrangements rather flexible in time and space. The post-industrial society has the potential to reestablish locational freedom similar to locational characteristics in the pre-industrial society. The individual search for quality of life will have greater impact on the whole settlement pattern in the post-industrial society.

Where, at the micro- as well as the macroregional level, demand for housing and services will grow is becoming less predictable. This favours some intermediate regions, while others will be losers. We anticipate seeing an increasing socioeconomic housing segregation in Sweden with clear microregional implications. A large number of households with less material and educational resources will have to cluster in housing areas in suburban and small towns with dense structure and poor services and physical environment.

The third driving force we emphasize is that the daily mobility of the population is reinforced by new transportation and communication technologies, permitting for a spatially wider range of activities for many individuals. Advanced combinations of private cars, public transportation and mobile telecommunications are changing and mixing the concepts of workplace, commuting, residence and service consumption. Among other things, this will lead to less long-distance migration. Changes in jobs will, to a decreasing degree, force people to migrate permanently. This allows for a more flexible housing market. More people are able and willing to look for places of residence further from production and service centres. We may anticipate that intermediate regions with good access to greater labour markets and with nice sites for private houses can provide attractive conditions. However, there will still be a large number of individuals that due to working conditions and personal resources have very limited mobility. This will contribute to the increasing housing segregation mentioned above.

The fourth driving force is policy-making and planning, where considerable changes are currently taking place. The evolution of the Swedish welfare state began in the 1930s with equalisation of living conditions in different social strata and regions a dominant goal. The main tools behind the development of the welfare system were large scale programs designed by central government. We now face a situation where Swedish policy-making is taking into consideration more market aspects and related rules of the game at the same time as there is decided to join the EU and adjust to supranational policy-making. This leads Swedish policy and planning towards the American model with much less public intervention, and much greater awareness and dependence on processes outside the borders of the national state.

The general effect of increased responsibilities of local government in Sweden will be to intensify competition, putting further stress not only on the comparative advantage of different municipalities, but also on management and entrepreneurship in local government. This stress is reinforced by the increasing financial problems generated by deficits in the national budget and constraints on the maximum level of local taxes. It is likely that this development will lead to a quantitatively and qualitatively more diversified and locally adjusted service production than before. Simultaneously, however, many municipalities/local governments will experience economic and social problems which will force them to call for more state resources and regulations. There is at present an opinion shift back to a stronger support for increasing resources to the public sector, which largely is a response to the present crisis at the labour market and in social services.

Not surprisingly, the shift towards more decentralisation is not free from conflicts. There is no doubt that Sweden is presently experiencing a crisis in municipal leadership, in rural, intermediate and urban municipalities, which is actually driving some municipalities towards the edge of bankruptcy.

The traditional top down character of the planning system has been criticised for lack of citizen participation. However, since the early 1980s many municipalities have been working on elaborating a more decentralised form of planning organisation involving citizen participation.

In summary, we expect the following spatial characteristics to emerge from the four carriers of change discussed above:

- a general trend of decentralisation of economic activity and political power with a stronger impact by both individual, commercial and local interests in regional development;
- a more uneven distribution of wealth and services between socioeconomic regions, discernible both at the micro- and macroregional level;
- a pressure towards profiling and exploitation of comparative advantages in each microregion, demanding a more entrepreneurial approach in local planning;
- a more complex system of linkages (economic, social, technological, political and cultural) within and between regions;
- marginality in terms of unemployment, under-employment, low disposable income level, ethnic and economic segregation, etc appears in all types of regions, metropolitan as well as rural.

As each region has a unique combination of various characteristics, i e different preconditions in terms of housing, production and network facilities, the consequences will be an increasingly visible divergence in the character between various intermediate regions. A new microregional fragmentation will appear.

7.3 Strategic Issues

We expect the following general problems to emerge from the fragmentation:

- more imbalances in terms of incomes and the quality and quantity of jobs and services;
- cumulative reinforcement of existing imbalances between regions, where strength now will lead to further strength in the future and vice versa;

and the following benefits:

- more local control promoting local entrepreneurship;
- increased possibilities of identifying and developing local niches;
- improved citizen participation in local and regional planning.

This move towards a new fragmentation is occupying governmental institutions responsible for national regional policy. In a decision by the Swedish Parliament of June 1994, based on Prop 1993/94:140, some principal changes in regional policy are implemented. The first is to increase the individual possibilities to choose job and place to live by contributing to development of well diversified regions all over the country. This is meant to be achieved by greater attention to the regional dimension in sectoral policies, including more locally adjusted organisational solutions, by more pronounced efforts to explore and exploit resources and growth potentials in rural areas, and by reinforced traditional regional policy in the special support areas.

The second is to take more consideration to differences in experiences, interests and competence between the sexes. Some new measures are introduced to back up female participation in regional development, especially as entrepreneurs.

Special emphasis is placed on the importance of developing higher formal qualifications among labour in the special regional support areas but also in other areas were the educational level is low. The needed increase in the number of educational slots will be spread to the smallest universities and colleges which simultaneously are provided with resources for research. It is also decided to put more attention on securing an acceptable standard of hard infrastructure, especially roads, in regions where lack of technical qualities needed for various types of industries exists. As the government earlier has decided upon a number of deregulations within different public services it is also stressed that these measures should not come in conflict with ambitions to secure an acceptable service standard in all regions.

It may be questioned if the strong and growing fragmentation tendencies may be neutralized by the regional policy measures decided upon. An aspect lacking in the new regional policy is how the important interplay between various types of regions should be handled. The spontaneous process of growing competition between regions within the country as well as across the national border simultaneously with creations of strategic alliances between the stronger regions will inevitably push towards further fragmentation and widened regional gaps in living standards. The critical transfers of resources between regions - the corner stone in the welfare model - will be much more difficult when unemployment is hitting all types of regions harder and in more complicated ways than ever before in modern time.

The present regional policy, largely oriented to specific regions in the northwestern parts of the country has to be replaced by a flexible system for stimulation of economic growth and distribution of welfare in any kind of microregion. Clearly, this is a major challenge to the efficient interplay between the supranational, the national and the local governments. Also, this is a major challenge to geographical research in order to monitor and analyze current economic and social processes at the microregional level. One task is to explore processes of marginalization in any type of region, irrespective of the Swedish "prejudice" that marginality is only found in remote and sparsely populated areas. There are better options than ever before in terms of available data bases, data processing capacity and illustrating techniques to identify various types of problems, follow processes of change, combine individual/household perspectives with macro perspectives and to evaluate efficiency of policy measures.

REFERENCES

Andersson, Å. E. (1986), The Four Logistical Revolutions. *Papers of the Regional Science Association*, 59, pp. 1 - 12.

Angel, D. P. and Mitchell, J. (1991), Intermetropolitan Wage Disparities and Industrial Change. *Economic Geography*, 67, pp. 124-35.

Axelsson, S., Berglund, S. and Persson, L. O. (1994), *Det tudelade kunskapssamhället*. Rapport 81, ERU, Stockholm.

Bengtsson, T. and Johansson, M. (1992), Population Structure, Community Development and Employment in Industrial Sweden. Paper presented at The First International Symposium on Population and the Comprehensive Development of Community, Hainan, China. Mimeo.

Berry, B. J. L. and Parr, J. J. (1988), *Market Centres and Retail Location: Theory and Applications*. Prentice-Hall, Englewood Cliffs.

Boltho, A. (1989), European and United States Regional Differentials: A Note. *Oxford Review of Economic Policy*, vol. 5, no. 2., pp 105-115.

Brotchie, J., Batty, M., Hall, P. and Newton, P. W., eds., (1991), *Cities of the 21st Century, New Technologies and Spatial Systems*. Longman Cheshire, Melbourne.

Bunte, R., Gaunitz, S. and Borgegård, L-E. (1982), *Vindeln. En norrländsk kommuns ekonomiska utveckling 1800-1980*. Lund.

Bylund, E. (1992), The Regional Level as Meeting Place, in M. Ó Cinnéide and S. Grimes, eds., *Planning and Development in Marginal Areas*. Centre for Development Studies, University College, Galway.

Bylund, E. (1994), Physical Planning According to the Planning and Building Law at Municipality Level in Sweden: An Attempt to Turn from Top-Down to Bottom-Up Planning?, in U. Wiberg, ed., *Marginal Areas in Developed Countries*. CERUM Report, Umeå University.

Carlsson, F., Johansson, M., Persson, L. O. and Tegsjö, B. (1993), *Creating Labour Market Areas and Employment Zones*. CERUM Report, Umeå University.

Castels, M. (1989), *The Informational City. Information, Technology, Economic Restructuring and the Urban-Regional Process*. Basil Blackwell.

Chang, J., Gade, O. and Jones, J. (1991) The Intermediate Socioeconomic Zone: a North Carolina Case Study, in O. Gade, V. P. Miller and L. Sommers, eds., *Planning Issues in Marginal Areas*. Appalachian State University, Boone, NC, Occasional Papers in Geography and Planning, Volume 3.

Clark, G. L. (1980), Capitalism and Regional Development. *Annals, Association of American Geographers*, 70, pp. 226-237.

Cox, K. R. and Mair, A. (1988), Locality and Community in the Politics of Local Economic Development. *Annals, Association of American Geographers*, 78, pp. 307-325.

Delors, J. (1989), Regional Implications of Economic and Monetary Integration, in *Report on Economic and Monetary Union in the European Community*. Luxemburg.

Drucker, P. F. (1993), *Post Capitalist Society*. Harper Collins, New York.

Ds 1992:81, *The Baltic Connection. Industrial Structure in and Integration between Southern Sweden and Northern Germany*. ERU, Stockholm.

Ds 1993:78, *En tillväxtfrämjande regionalpolitik*. Ministry of Labour (1993), Stockholm.

Ennefors, K. (1991), *Att leva mellan stad och land - livskvalitet i den urbaniserade glesbygden*. CWP 1991:15, CERUM, Umeå University.

Fisher, A. G. B. (1933), Capital and the Growth of Knowledge. *Economic Journal*, vd. XLIII.

Fisher, A. G. B. (1939), Production: Primary, Secondary and Tertiary. *Economic Record*.

Friedmann, J. (1987), *Planning in the Public Domain*. Princeton University Press. Princeton NJ.

Gade, O. (1991), Functional and Spatial Problems of Rural Development: General Considerations of an Exploratory Model, Paper presented at the International School of Rural Development, Galway University College, Ireland.

Gade, O., Persson, L. O. and Wiberg, U. (1992), Processes Shaping the Rural-Urban Continuum: The Cases of Sweden and North Carolina, USA, in M. Ó Cinnéide and S. Grimes, eds., *Planning and Development in Marginal Areas*. Centre for Development Studies, University College, Galway.

Goldthorpe, J. H., Lockwood, D., Bechhofer, F. and Platt, J. (1968), *The Affluent Worker: Political Attitudes and Behaviour*. Cambridge University Press.

Grotewold, A. (1959), Von Thünen in Retrospect. *Economic Geography*, 35, p. 350.

Hall, P. (1990), Urban Europe after 1992, in *Urban Challenges*. SOU 1990:33, Stockholm.

Hall, P. and Cheshire, P. C. (1988), *Growth Centres in the European Urban System.* Heinemann Education, London.

Hansen, N. M. (1970), Rural Poverty and the Urban Crisis: *A Strategy for Regional Development.* Indiana University Press, Bloomington.

Henrekson, M. (1993), Humankapital, produktivitet och tillväxt, i *Nya villkor för ekonomi och politik.* SOU 1993:16, Bilagedel 1, Stockholm.

Holmberg, S. and Gilliam, M (1989), *Sverige: Från klassväljare till åsiktsröstare.* Department of Political Science, Gothenburg University.

Häggroth, S. (1993), *From Corporation to Political Enterprise. Trends in Swedish Local Government.* Ds 1993:6, Ministry of Public Administration, Stockholm.

Höjrup, T. (1983), *Det glemte folk. Livsformer og centraldirigering.* Institut for Europeisk Folkelivsforskning og Statens Byggeforskningsinstitut, Copenhagen.

Inglehart, R. (1979), Werthandel und politisches verhalten, in J. Matthes, ed., *Sozialer wandel in Westeuropa.* Frankfurt.

Jackson, R. W., Hewings, J. H. and Sonis, M. (1989), Decomposition Approaches to the Identification of Change in Regional Economics. *Economic Geography*, 65, pp. 216-231.

Johannisson, B., Persson, L. O. and Wiberg, U. (1989), *Urbaniserad glesbygd - verklighet och vision.* Ministry of Labour, Ds 1989:22, Stockholm.

Johansson, B. (1989), *Economic Development and Networks for Spatial Interaction.* CWP 1989:28, CERUM, Umeå University.

Johansson, B. and Karlsson, C. (1990), Evolving Technological Patterns in a Nordic Perspective, in B. Johansson and C. Karlsson, *Innovation, Industrial Knowledge and Trade - A Nordic Perspective.* European Networks 1990:1, CERUM, Umeå University and Institute for Futures Studies, Stockholm.

Johansson, B. and Karlsson, C. (1991), Technology Development and Regional Infrastructure. Revised version of paper presented at the 38th NARSA-meeting in New Orleans, November 1991.

Johansson, M. and Persson L. O. (1991), *Regioner för generationer.* Allmänna Förlaget, Stockholm.

Jones, J. (1991), *A Comparative Analysis of the Intermediate Socioeconomic Zone in Sweden and North Carolina.* Appalachian State University, Boone, North Carolina. Master's Thesis.

Keeble, D. (1991), High Technology Industry and the Restructuring of the UK Space Economy, in P. Townroe and R. L. Martin, eds., *Regional Development in the 1990s: Britain and Ireland in Transition.* Jessica Kingsley, London.

Korpi, W. (1978), *Arbetarklassen i välfärdskapitalismen. Arbete, fackförening och politik i Sverige.* Prisma, Stockholm.

Larsson, J. (1992), Growth of the Knowledge Society in a Regional Perspective, in L. Andersson and C. Karlsson, eds., *The Medium-Sized City. Research for Renewal.* Research Report 92:3, University of Karlstad.

Lipset, S. M. (1986), North American Labor Movements: A Comparative Perspective, in S. M. Lipset, ed., *Unions in Transition. Entering the Second Century.* ICS Press, San Fransisco.

Millan, B. (1993), Memorandum to the Informal Council of Ministers responsible for Regional Policy and Spatial Planning. Nov. 1993, mimeo.

National Atlas of Sweden (1991), *The Population.*

Nielsen, H. J. (1990), *American Individualism and American Labour Union Decline.* Research Report, Institute of Political Studies, University of Copenhagen.

NUTEK (1994), *Statsbudgetens regionala fördelning.* R 1994:3, Stockholm.

OECD (1991), *Economic Surveys: Sweden.* Paris.

OECD (1993), *From Higher Education to Employment.* Synthesis report, Paris.

Ohlsson, L. and Vinell, L. (1987), *Tillväxtens drivkrafter. En studie av industrins framtidsvillkor.* Industriförbundets förlag, Stockholm.

Parr, J. B. (1987), Interaction in an Urban System: Aspects of Trade and Commuting. *Economic Geography*, 63, pp. 223-240.

Persson, L. O. (1991), Urbanizing Processes in Peripheral Areas in a Welfare State, in O. Gade, V. P. Miller and L. Sommers, eds., *Planning Issues in Marginal Areas.* Appalachian State University, Boone, NC, Occasional Papers in Geography and Planning, Volume 3.

Persson, L. O. (1992), Changing Functions of Regions Beyond the Metropolitan Edge, in O. Gade, ed., *Spatial Dynamics of Highland and High Latitude Environments.* Appalachian State University, Boone, NC, Occasional Papers in Geography and Planning, Volume 4.

Persson, L. O. and Wiberg, U. (1988), Policy and Planning for the Urbanized Rural Areas. *Scandinavian Housing and Planning*, Vol 5:4.

Phillips, K. (1990), *The Politics of Rich and Poor: Wealth and the American Electorate in the Reagan Aftermath.* Random House, New York.

Piatier, A. (1981), Innovation, Information and Long Term Growth. *Futures* 5.

Prop 1993/94:140, *Bygder och regioner i utveckling.* Policy document from the Swedish Government.

Reich, R. B. (1992), *The Work of Nations.* Vintage Books, New York.

Reiman, R. E. (1992), Bottom-Up Planning in a Marginal Area of Southeastern North Carolina, USA: A Case Study, in M. Ó Cinnéide and S. Grimes, eds., *Planning and Development in Marginal Areas.* Centre for Development Studies, University College, Galway.

Roepke, H. G. and Freudenberg, D. A. (1981), The Employment Structures of Nonmetropolitan Counties. *Annals, Association of American Geographers*, 71, pp. 580-592.

Schor, J. B. (1992), *The Overworked American*. Basic Books, New York.

Sjöholt, P. (1990), Producer Services: A New Panacea for Marginal Regions, in L. Lundqvist and L. O. Persson, eds., *Nätverk i Norden*. Ds 1990:78, Stockholm.

Socialstyrelsen (1994), *Social Rapport 1994*. Stockholm.

SOU 1989:65, *Staten i geografin*. Ministry of Labour, Stockholm.

SOU 1992:19, *Långtidsutredningen*. Ministry of Finance, Stockholm.

Stöhr, W. B. (1974), *Interurban Systems and Regional Economic Development*. Association of American Geographers, Comm. on College Geogr., Resource paper No 26, Washington DC.

Suarez-Villa, L. and Cuadrado Roura, J. R. (1993), Regional Economic Integration and the Evolution of Disparities. *Papers in Regional Science: The Journal of the RSAI*, vol. 72, no. 4, pp. 369-388.

Swedish Government's Financial Plan 1991/92.

Swedish Productivity Delegation, 1991.

The Prime Ministers' Declaration to Parliament, 1991.

Thurow, L. (1992), *Head to Head. The Coming Economic Battle Among Japan, Europe and America*. Morrow, New York.

Törnqvist, G. (1988), *System of Cities in Changing Technical Environment*. CWP 1988:12, CERUM, Umeå University.

Veggeland, N. (1990), The Circular Motion of Planning. *Nordisk Samhällsgeografisk Tidskrift*, 12.

Veggeland, N. (1991), Norge i et regionenes Europa. *Stat og styring*, No 5, Oslo.

Veggeland, N. (1993), Rationalism or Federalism? Two Visions of a New Europe, in L. Lundqvist and L. O. Persson, eds., *Visions and Strategies in European Integration. A North European Perspective*. Springer-Verlag, Berlin-Heidelberg.

Viatte, G. and Cahill, C. (1991), The Resistance of Agricultural Reform. *The OECD Observer Aug/Sept 1991*.

Wiberg, U., ed. (1990) *Characteristics of the Intermediate Zone in Sweden and USA*. CWP 1990:8, CERUM, Umeå University.

Wiberg, U. (1992a), Implications for Regions Beyond the Edge of Shifts in Policy and Planning in Sweden, in O. Gade, ed., *Spatial Dynamics of Highland and High Latitude Environments*. Appalachian State University, Boone, NC, Occasional Papers in Geography and Planning, Volume 4.

Wiberg, U. (1992b), The Role of Urban Centres in Fringe Areas, in M. Tykkyläinen, ed., *Development Issues and Strategies in the New Europe*. Avebury, Aldershot.

Wiberg, U. (1994a), Swedish Marginal Regions and Public Sector Transformation, in U. Wiberg, ed., *Marginal Areas in Developed Countries*. CERUM Report, Umeå University.

Wiberg, U. (1994b), Marginal Areas and Policy-Making in Sweden, in C-Y. D. Chang, ed., *Marginality and Development Issues in Marginal Regions*. Proceedings of the Study Group on Development Issues in Marginal Regions, International Geographical Union, Taipei, Taiwan.

Wilson, D. (1991), Urban Change, Circuits of Capital, and Uneven Development. *Professional Geographer*, 43, pp. 403-415.

Worre, P. (1982), Class Parties and Class Voting in Scandinavian Countries. *Scandinavian Political Studies*, Vol. 3, New Series, pp. 299-320.

AUTHOR INDEX

Andersson, Å. E.	15		Gade, O.	1, 5, 19, 44, 88, 106
Angel, D. P.	6			
Axelsson, S.	51, 68-70, 73-84		Gaunitz, S.	14
			Gilliam, M.	20
Batty, M.	23		Goldthorpe, J. H.	20
Bechhofer, F.	20		Grotewold, A.	3
Bengtsson, T.	72		Hall, P.	16, 22-23
Berglund, S.	51, 68-70, 73-84		Hansen, N. M.	24
			Henrekson, M.	61
Berry, B. J. L.	4		Hewings, J. H.	5
Boltho, A.	97		Holmberg, S.	20, 110
Brotchie, J.	23		Häggroth, S.	102, 104
Bunte, R.	14		Höjrup, T.	21
Borgegård, L-E.	14		Inglehart, R.	20
Bylund, E.	103		Jackson, R. W.	5
Cahill, C.	109		Johannisson, B.	47-48, 92
Carlsson, F.	45, 63-64, 66		Johansson, B.	15-16, 18, 92
Castels, M.	17		Johansson, M.	45, 47, 54, 63-64, 66, 72
Chang, J.	5, 88			
Cheshire, P. C.	22		Jones J.	5, 88, 90, 93
Clark G. L.	6		Karlsson, C.	16, 18, 92
Cox K. R.	6		Keeble, D.	23
Cuadrado Roura, J. R.	97		Korpi, W.	20
Delors, J.	97		Larsson, J.	26
Drucker, P. F.	61		Lipset, S. M.	43
Ennefors, K.	37		Lockwood, D.	20
Fisher, A. G. B.	13		Mair, A.	6
Freudenberg, D. A.	24		Millan, B.	116
Friedmann J.	105		Mitchell, J.	6
			Newton, P. W.	23

Nielsen, H. J.	21
Ohlsson, L.	27
Parr, J. B.	4
Parr, J. J.	4
Persson, L. O.	1, 19, 45, 47-48, 51, 54, 63-64, 66, 68-70, 73-84, 92, 106
Philips, K.	43
Piatier, A.	25
Platt, J.	20
Reich, R. B.	58-60
Reiman, R. E.	108
Roepke, H. G.	24
Schor, J. B.	61
Sjöholt, P.	47
Sonis, M.	5
Stöhr, W. B.	106-107
Suarez-Villa, L.	97
Tegsjö, B.	45, 63-64, 66
Thurow, L.	52
Törnqvist, G.	16
Veggeland, N.	97, 105
Viatte, G.	109
Wiberg, U.	1, 19, 24, 47-48, 92, 106
Wilson, D.	6
Vinell, L.	27
Worre, P.	20

SUBJECT INDEX

accessibility	3, 6, 18, 30, 60, 112, 116, 117	housing	5, 6, 15, 21, 40, 45, 76, 85, 89, 100, 102, 111, 118
attractivity	33, 114		
bottom-up	60, 103, 108, 112	image	49, 92, 104, 108, 115, 118
central place theory	4, 102		
cluster analysis	24, 45, 64, 94	immigrants	8, 89, 118
collectivism	18, 49, 100	income distribution	45
commuting	14, 23, 31, 33, 42, 45, 55, 62, 85, 88, 118	individualism	1, 18, 43, 49
		information society	15
		information technology	15, 23, 30
competitiveness	2, 16, 52, 60, 99, 109	infrastructure	15, 22, 30, 38, 47, 95, 96, 101, 103, 106, 111, 115, 117, 121
decentralisation	1, 19, 101, 104, 119		
		intermediate regions	13, 24, 27, 33, 47, 65, 89, 104, 116, 118
deindustrialisation	15, 22, 69		
deregulation	10, 19		
education	5, 6, 18, 27, 40, 47, 51, 58, 63, 71, 72, 85, 88, 100, 107, 118	internationalisation	25, 47, 117
		ISER (Intermediate socio-economic region)	2, 6, 16, 19, 24, 94, 101, 107
		knowledge society	15, 51, 58, 62
European integration	6, 87, 96, 110	life style	19, 32, 117
European Union (EU)	96, 109, 119	local labour market	7, 16, 27, 33, 40, 45, 51, 62, 67, 73, 85, 99, 115
foreign investment	28		
fragmentation	67, 113, 115, 117, 120		
globalization	1, 25	local planning	104, 120
goods-handling	8, 22, 25	macroregional	42, 115, 118
		market orientation	16

medium-sized cities	22, 24, 38, 42, 58, 101	public services	39, 45, 67, 72, 85, 89, 100, 103, 121
metropolitan	3, 6, 21, 22, 24, 27, 40, 45, 53, 65, 67, 75, 87, 88, 104, 106, 116, 120	qualification	45, 51, 58, 63, 80, 111, 121
		quality of life	1, 3, 6, 18, 23, 37, 49, 91, 107, 118
microregional	27, 115, 117, 121	regional planning	92, 105, 106, 120
migration	33, 45, 60, 72, 85, 118	regional policy	3, 15, 40, 55, 62, 67, 91, 100, 103, 115, 120
mobility	2, 8, 17, 31, 33, 45, 73, 85, 87, 99, 106, 117	service economy	96
mosaic	97	single european market	96
municipality	14, 33, 45, 55, 88, 99, 102, 103, 111	small towns	24, 33, 76, 85, 118
		socio-economic continuum	5, 88, 99, 106
national state	115, 119	solidarity	20, 49, 97, 100
network	4, 13, 25, 60, 95, 117	spatial structure	13, 22
North Carolina	32, 44, 88, 108	subcontracting	6
periphery	4, 10, 27, 33, 40, 47, 56, 87, 94, 97, 107	supranational	119, 121
		telecommunication	15, 24, 30, 47, 97, 100, 117
place-hunting	16	top-down	103
postmaterialistic	19	trade barriers	115
postindustrial	46	transfers	7, 40, 87, 92, 103, 121
poverty level	9		
preferences	13, 16, 23, 33, 46, 81, 87	urbanization	44, 88, 102
		urbanized rural area	47, 92
privatisation	1, 19, 112	welfare state	3, 7, 19, 87, 99, 107, 109, 117
productivity	51, 61, 100, 109		
public expenditures	39		
public sector	6, 14, 33, 39, 55, 62, 63, 87, 100, 103, 109, 115, 118		